Contents

Text and illustration credits

We are grateful to the following for permission to reproduce copyright material.

Gulf Publishing Co. for the article 'Evaluate risk in plant design' by Trevor Kletz from *Hydrocarbon Processing* May 1977. Copyright 1977 by Gulf Publishing Co., all rights reserved; Macmillan Magazines Ltd and the author, P E Bryant, for the article 'Arithmetic in the cradle' from *Nature* Vol. 358, pp. 712–713, Copyright © 1992 Macmillan Magazines Ltd; IPC Magazines Ltd for the articles 'The formula man' by Ian Stewart pp. 24–28, 18.12.87, 'Geometry, shoemaking and the Milk Tray problem' by Ray Cuninghame-Green pp. 50–53, 15.7.92 and 'When the turning gets tough ...' by Kerry Spackman and Sze Tan pp. 28–31, 13.3.93, from *New Scientist*; the author, Adrian Oldknow, from the Mathematics Centre, West Sussex Institute of Higher Education, for his article 'A tale of two cheeses'; Transport Research Laboratory for the article 'Traffic modelling and statistics' by Mike Maher, which originally appeared in *Teaching Statistics* Vol. 14 No. 2 1992, Crown Copyright 1992.

We are grateful to the following for permission to reproduce copyright material.

Trevor Kletz, ICI, from 'Evaluate risk in plant design', *Hydrocarbon Processing*, May 1977, Figures 8.1 and 8.2; *New Scientist*, Kerry Spackman and Sze Tan, 'When the turning gets tough ...', March 1993, Figures 9.1–9.8, Ian Stewart, 'The formula man', December 1987, Figures 5.1, 5.2 and 5.3, Ray Cuninghame-Green, 'Geometry, shoemaking and the Milk Tray problem', August 1989, Figures 5.1–5.7; Transport Research Laboratory, Mike Maher, *Teaching Statistics* Vol. 14 No. 2 Summer 1992, 'Traffic modelling and statistics', Figure 1.1 and 1.2.

Photograph of Florence Nightingale reproduced by permission of the Mansell Collection. Photograph from 'Traffic modelling and statistics' reproduced by permission of the Transport Research Laboratory.

Longman Group UK Limited
Longman House, Burnt Mill, Harlow, Essex CM20 2JE,
England and Associated Companies throughout the World.

© Nuffield Foundation 1994

First published 1994
ISBN 0582 09989 7

Set in 10/12 Times by Nina Towndrow
Illustrated by Hugh Neill and Oxford Illustrators Limited
Produced by Longman Singapore Pte Ltd
Printed in Singapore

The publisher's policy is to use paper manufactured from sustainable forests.

1 *Traffic modelling and statistics*

Mike Maher

This article illustrates some aspects of the role of mathematical and statistical modelling in the study of traffic. Although the article describes two models which are used in traffic planning, it can not give the details which would be found in a book about traffic modelling. The important idea is that the models themselves suggest areas of research and development which might be profitable in improving road conditions or in reducing the number of accidents.

You can read more about the ideas used in traffic modelling in books such as *Highways Volume One - Traffic Planning and Engineering* by C A O'Flaherty, published by Edward Arnold.

Introduction

The Transport and Research Laboratory (TRL), based at Crowthorne in Berkshire, is part of the Scientific Civil Service, and is responsible for carrying out a large part of the research programme of the government's Department of Transport. The range of the research activities of the laboratory is wide, covering such disparate topics as: the structural assessment of bridges, the effectiveness of road safety publicity, vehicle impact studies, and road construction techniques for developing countries. The 600 or so scientific and support staff at TRL are drawn from a variety of disciplines: physics, engineering, geology, computer science, geography and mathematics. The highest proportion of mathematicians and statisticians is to be found in Traffic Group, where the work involves a great deal of mathematical and computer model development and testing.

Most parts of the laboratory's research programme involve empirical studies of one sort or another, so that the analysis of data is a frequent and important activity. Whilst much of this analysis is quite simple and routine, and can be carried out by the staff concerned with the aid perhaps of statistical packages, some of it requires specialist treatment and the development of particular methodologies. Because of my statistical background, I therefore sometimes act as a statistical consultant, giving advice or reassurance to staff working on a variety of projects. For the most part, however, I work on the research programme of the Traffic Safety Division, operating as a traffic modeller rather than as a statistician per se.

The underlying aim of TRL's work is to carry out studies which will produce results, or to develop tools, which will be used, typically, by the local authority engineer or planner. For example, the objective might be to develop computer models of traffic flow, at junctions or through road networks, which can be used in the economic and operational appraisal of schemes such as by-passes: what effect will it have?; how much traffic will use the by-pass?; how much congestion in the town centre will be relieved?; and is the capital expenditure justifiable? Models are therefore required to describe drivers' route choices, and to predict the capacity of, and the delays produced by, a junction of any prescribed design. Other models are required to predict the safety consequences of the re-distribution of flows through a network, or a change in the design of a junction. Such models can only be developed through empirical studies, by identifying those factors which have a significant effect (on delay or accident rate, for example), and by establishing the nature and magnitude of the relationship.

Data collection, data analysis and model building are, then, vital and standard parts of the research work of TRL. However, in order to give a flavour of my work, it will be better if I move from the general to the particular, and describe some examples of pieces of research, in which statistics play an important role. The examples I have chosen are taken from the work of my own Division, Traffic Safety.

Accident models

The number of accidents at a site (such as a roundabout) is likely to be influenced by the flows passing through the site and the physical characteristics, or design, of the site, some of which it may be possible to alter. To establish these effects, there are two alternative approaches: cross-sectional studies and longitudinal studies. Let us look in turn at each of these, and then compare them.

Cross-sectional studies

In a cross-sectional study, a suitably stratified sample of sites is taken and data are collected on the accident frequency, the flows and those variables which might be thought to exert an effect on the accident frequency. After careful regression model building and testing, it is possible to establish a model which, on the one hand, is as simple as possible, but on the other contains all the explanatory variables which exert a significant influence on accident rate. For example, it has been found that the average number of accidents per year at a roundabout is given by:

$$\mu = 0.0454 Q^{1.22} \qquad (1)$$

in which Q is the total daily flow entering the roundabout (expressed in thousands). Figure 1.1 shows data from a study of 78 roundabouts, together with the curve showing the above model.

For more detailed modelling, the data can be disaggregated into separate arms of the junction and separate accident type. For example, the average number of accidents per year on an approach to a roundabout has been found to be given by:

$$\mu = 0.0057 Q^{1.7} \exp(20C - 0.1w) \qquad (2)$$

in which Q is the daily entering flow (in thousands) on that approach, C is the entry curvature (in metres^{-1}), and w is the entry width (in metres). Models of similar structure have been developed for other accident types, and for other junction types. Armed with such models (and other models which predict the capacity and delays), the engineer is then in a position to explore the consequences, in terms of delay and safety, of any particular design. He could even consider the effects of changing from a roundabout to signals.

Figure 1.1

Accidents versus entry flow at roundabouts

In the above model, it can be seen that the effect of increasing the entry width is to reduce the accident rate. In fact, it predicts that increasing the entry width by one metre has the effect of reducing the accident rate by approaximtely 10%. The relationship established between accident rate and entry width is, strictly, purely associative and not necessarily causative. To be able to make such inferences about the effect of changing the design of a junction, the engineer needs to be able to treat it as if it **were** causative. It is interesting to consider under what conditions such an assumption is justifiable. In an attempt to avoid such problems, the alternative approach of before-and-after studies might seem attractive.

Longitudinal studies

In a before-and-after investigation, the treatment (such as the installation of a pedestrian crossing, or increasing the entry width by one metre) is applied at a number of sites and the accident rate in the after period compared with that in the before period. Unlike the cross-sectional approach, the effect is inferred directly. There are disadvantages, however: one must know in advance which treatment is to be investigated, and one must wait until the after period is complete before results can be obtained. All other factors should remain unchanged, so that any change in accident rate may be safely attributed to the treatment. Now, unfortunately for the scientist, traffic safety research is not a laboratory-based subject, and the clean and purely designed and controlled experiment rarely exists. Data is often obtained through the cooperation of local authorities, and political, financial, ethical and other practical influences may come into play. For example: why and how are the study sites selected for treatment? Perhaps they tend to be those felt to be accident "blackspots"; that is, those with a poor accident record in the before period.

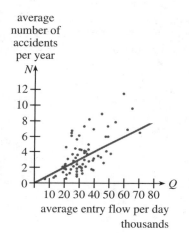

average number of accidents per year

N

12
10
8
6
4
2
0

10 20 30 40 50 60 70 80 $\quad Q$

average entry flow per day

thousands

Figure 1.2

Apparent reduction in accident rate following "treatment"

The effect of such a selection criterion is easy to see. Take the 78 roundabouts considered earlier, with their underlying accident rates given by the model in (1), let us assume. Then, imagine the accident frequencies which might occur in one year by generating Poisson variates with the above set of means. Suppose that the sites selected for "treatment" were those with the highest accident frequencies in the "before" period, but that the "treatment" left the true mean accident rates unaffected in the "after" year. Then the estimate of the effectiveness of the treatment, obtained from the comparison of before and after accident frequencies, would be likely to be as shown in Figure 1.2, from which it can be seen that, if a small proportion of sites are selected for treatment, the estimate of the effectiveness can reach 30%, **even though no treatment is applied**.

(This phenomenon is easily repeated, by generating random variates from a gamma distribution with, say, mean 3 and a shape parameter of 4, to represent the true means, and then generating Poisson variates with those means, using MINITAB.)

It is clear, then, that considerable bias can be introduced by the process of selection of sites. Another potential problem is that other conditions rarely remain constant during the course of the before-and-after study: traffic flows may increase, for example. The standard way to allow for such extaneous effects is through the use of a group of **control sites**. The role of the controls is then to permit the estimation of what would have happened, on average, at the study sites if no treatment had been applied. The introduction of the control observations into the analysis inevitably adds uncertainty into the estimation: the larger the control group, the smaller the additional amount of uncertainty. Ideally, the control group should be "similar" to the study group. However, it turns out to be surprisingly difficult in practice to identify a good control.

Conclusions

These then are some of the problems which the statistician may have to face in the analysis of what at first sight appears to be a fairly straightforward set of data. Largely because the data come from the "real world", with all its imperfections and with the inevitable conflict between research needs and practical application, simple textbook statistical techniques are sometimes not valid. The statistician working on traffic modelling must find the right balance between rigour and applicability, develop a feel for how far the simple techniques can be stretched, and be capable of developing new methodologies when necessary. The rewards come from working on an inexhaustible supply of challenging and interesting problems, having the satisfaction of seeing the results implemented in "real world" applications, and helping others obtain sound interpretations of their results.

2 The formula man

Ian Stewart

1987 celebrated the centenary of the birth of Srinivasa Ramanujan. He was a self-taught mathematician who, born into poverty, wrote his formulas on used wrapping paper – until a Cambridge don recognised the originality of his work.

The story of Ramanujan's short life is remarkable by any criteria – it is both a triumph and a tragedy. But as you read the story, think about how Ramanujan might have discovered his formulae, how he tested them, improved them, gradually became certain that they were correct and then attempted to justify them.

In some ways the work that Ramanujan did has parallels in the *Investigating and proving* unit in Book 4.

It was January 1913. Turkey was at war in the Balkans and Europe was being dragged deeper and deeper into the conflict. Godfrey Harold Hardy, a mathematician at Cambridge University, despised war. He took great pride that the field of his life's work, pure mathematics, had no military uses. "Real mathematics has no effect on war," he later wrote in his book *A Mathematician's Apology*. Innocent times.

Outside, snow drizzled damply down, while begowned undergraduates scurried through the slush of Trinity Great Court. But in Hardy's rooms a cheerful fire kept the cold at bay. On the table lay the morning post, ready to be opened. Hardy glanced at the envelopes. One caught his eye because of its unusual postage stamps.

India. Postmarked Madras, 16 January 1913.

Hardy slit the heavy manila envelope, more than a little battered by its long journey, and drew out a sheaf of papers (see Figure 2.1 overleaf).

An accompanying letter, in an unfamiliar hand, began:

> Dear Sir,
>
> I beg to introduce myself to you as a clerk in the Accounts Department of the Port Trust Office at Madras at a salary of only £20 per annum. I am now about 23 years of age. I have had no University education ... After leaving school I have been employing the spare time at my disposal to work at Mathematics ... I am striking out a new path for myself.

Oh Lord, another crank. Probably thinks he's squared the circle. Hardy nearly put the letter aside. But a sheet of mathematical symbols caught his eye. Curious formulas. A few, he recognised. Others were ... unusual.

If the author of the letter is a crank, he might at least prove an entertaining crank. Hardy read on.

> Very recently I came across a tract by you styled Orders of Infinity in page 36 of which I find a statement that no definite expression has as yet been found for the number of prime numbers less than any given number. I have found an expression which very nearly approximates the real result, the error being negligible.

My word. He's reinvented the Prime Number Theorem.

> ... I would request you to go through the enclosed papers. Being poor, if you are convinced there is anything of value I would like to have my theorems published ... Being inexperienced I would very highly value any advice you give me. Requesting to be excused for the trouble I give you.

> I remain, Dear Sir, Yours truly, S. Ramanujan

Not a typical crank, Hardy mused. *A typical crank would be more aggressive and more conceited.* He put the letter aside and picked up the *Times* newspaper, to read the Australian cricket scores. Then it was time for his lecture on analysis. He shrugged into his gown, walked out of the room and shut the door behind him.

As was his habit, he reserved the morning for mathematics, and spent the afternoon playing real tennis at the court near the university library. But, as he strolled back to his rooms, the morning's events continued to play on his mind. Some wildly improbable formulas, plus some very familiar ones, presented as if they were new – a discomforting mixture.

That evening, over High Table, he described the curious turn of events to any of the fellows of the college who cared to listen, including his colleague and close collaborator, John Littlewood. Port and walnuts in the common room put Littlewood in a mellow mood, and he was willing to waste an hour to help put his friend's mind at ease. The chess room was free. As they entered the room, Hardy held up the slim sheaf of paper. "This man", he said to the gathering at large, "is either a crank or a genius."

An hour later, Hardy and Littlewood emerged with the verdict. *Genius.*

We can reconstruct Hardy's movements, with only a little poetic licence, from C. P. Snow's introduction to *A Mathematician's Apology*. As he says there, Hardy always followed the same routine. We also know a lot about Ramanujan, thanks to his biographers P. V. Seshu Aiyar and Ramachandra Rao. Hardy and Littlewood also put various reminiscences on record.

A mathematician, natural-born

Srinivasa Ramanujan Aiyangar was born into a Brahmin family on 22 December 1887. His father, carrying on a family tradition, was a minor accountant; his mother was a bailiff's daughter. The birth took place in his grandmother's house in Erode, a town in the southern province of Tamil Nadu, India. He grew up in Kumbakonam, where his father worked. Kumbakonam is 160 miles south-west of Madras, and Erode is 120 miles to the west of Kumbakonam. The family was poor, the house tiny, but it was a happy childhood.

Ramanujan's mathematical talents surfaced early. At the age of 12, he borrowed an advanced textbook and mastered it completely, without apparent effort. His biographers record one incident:

$$\frac{1}{2^4} + \frac{1}{3^4} + \frac{1}{5^4} + \frac{1}{7^4} + \ldots = \frac{15}{2\pi^4}$$

where the sum is over all numbers 2, 3, 4, 5, ... with an odd number of distinct prime divisors.

$$\frac{\coth \pi}{1^7} + \frac{\coth 2\pi}{2^7} + \frac{\coth 3\pi}{3^7}$$
$$+ \ldots = \frac{19\pi^7}{56\,700}$$

$$\int_0^a e^{-x^2}\, dx = \frac{\sqrt{\pi}}{2} - \frac{e^{-a^2}}{2a}$$
$$+ \frac{1}{a+} \frac{2}{2a+} \frac{3}{a+} \frac{4}{2a+} \ldots$$

$$\frac{1}{1+} \frac{e^{-2\pi}}{1+} \frac{e^{-4\pi}}{1+} \frac{e^{-6\pi}}{1+} \ldots =$$

$$\left(\sqrt{\frac{5+\sqrt{5}}{2}} - \frac{\sqrt{5}+1}{2} \right) 5\sqrt{e^{2\pi}} e^{\pi}$$

The last two formulas are continued fractions: the notation
$$\frac{1}{1+} \frac{a}{1+} \frac{b}{1+} \frac{c}{1+} \ldots$$
means
$$\cfrac{1}{1+\cfrac{a}{1+\cfrac{b}{1+\cfrac{c}{\ddots}}}}$$

Figure 2.1

Some of the formulas Ramanujan sent to Hardy

> "While he was in the second form he had it appears, a great curiosity to know the 'highest truth' in Mathematics, and asked some of his friends in the higher classes about it. It seems that some mentioned the Theorem of Pythagoras as the highest truth, and that some others gave the highest place to 'Stocks and shares'."

At the age of 15 an event occurred that was to change his life; but at the time, it seemed mundane rather than momentous. He managed to borrow a copy of G. S. Carr's *Synopsis of Elementary Results in Pure Mathematics* from the government college library.

The *Synopsis* is idiosyncratic. It lists some 6000 theorems – without proofs. Carr based it on problems that he set when coaching students. Ramanujan likewise set himself a problem: to establish *all the formulas in the book*. He had no help, no other books. Effectively he had set himself a research project of 6000 separate topics. Ramanujan devised his own methods, and wrote down his discoveries in the first of a series of notebooks, which he kept throughout his life.

Too poor to afford paper, he did his calculations on a slate and jotted the results in a series of notebooks. In 1909, he married. In February 1912, Ramanujan's mathematics lecturer, P. V. Seshu Aiyar, sent him to see R. Ramachandra Rao, then collector at Nellore. Rao's record of the interview includes the following:

> "… I condescended to permit Ramanujan to enter my presence. A short uncouth figure, stout, unshaven, not over-clean, with one conspicuous feature – shining eyes … I saw quite at once that there was something out of the way; but my knowledge did not permit me to judge whether he talked sense or nonsense … He showed me some of his simpler results. These transcended existing books and I had no doubt that he was a remarkable man. Then, step by step, he led me to elliptic integrals and hypergeometric series and at last his theory of divergent series not yet announced to the world converted me."

Rao secured Ramanujan an appointment in the Madras Port Trust Office, at 30 rupees per month, a job that left him enough spare time to continue his research. Another advantage was that he could take away used wrapping paper, to write his mathematics on.

It was then that, at the urgings of the same people, Ramanujan wrote his diffident letter to Hardy.

Hardy wasted no time. Convinced that the Indian clerk was a natural-born mathematician of the highest order, he immediately sent an encouraging reply. Ramanujan lost no time in writing back:

> … I have found a friend in you who views my labours sympathetically … Verify the results I give and if they agree with your results … you should at least grant that there may be some truths in my fundamental basis … To preserve my brains I want food and this is now my first consideration. Any sympathetic letter from you will be helpful to me here to get a scholarship either from the University or the Government.

But Hardy had already made his move. He had written to the secretary for Indian students in London, to ask whether there was some way to get Ramanujan a Cambridge education. It then turned out that Ramanujan did not want to leave India. By way of the University of Madras, this decision was transmitted to Hardy.

Not to be defeated, the Cambridge network rolled into action. Another Trinity mathematician, G. T. Walker, was visiting Madras at the time, and he was also apprised of Ramanujan's work. He wrote a letter to the University of Madras, and on that basis, the university granted a special scholarship. For the first time, Ramanujan was free to devote all of his time to mathematics.

Hardy did not give up. He continued trying to persuade Ramanujan to come to England, through letters, and through the personal intervention of another Trinity Fellow, E. H. Neville. Ramanujan began to waver: the main obstacle became his mother. Then, one morning and to general family astonishment, the mother announced that the goddess Namagiri had appeared to her in a dream, commanding her not to prevent her son fulfilling his life's calling.

Ramanujan was provided with a grant to cover subsistence for two years, plus free travel, and sailed for England. By April 1914, he was in Trinity College. He must have felt very out of place at Trinity, but he stuck to it, publishing many research papers, including some very influential joint work with Hardy.

But many stumble on the path to glory. In 1917, Ramanujan's health began to fail, and he went into a nursing home. In the meantime, he had been elected a Fellow of the Royal Society, the first Indian to be so honoured. Trinity elected him to a fellowship.

Hardy and Littlewood come out of the story remarkably well: others do not. Ramanujan had previously approached two other prominent English mathematicians, who returned his manuscripts without comment. Hardy and Littlewood also played a prominent role in the academic politics of the Trinity fellowship. In his book, *A Mathematician's Miscellany*, Littlewood says: "I am the only person who knows the facts, and they should be put on record, if only as illustrating the fantastic state of the College just after the 1914-1918 war …"

Several people opposed Ramanujan's election, one apparently out of colour prejudice. The fact that Ramanujan already had an FRS weighed heavily, but the opposition saw it as a dirty trick. Hardy was not on the election committee but Littlewood was, and fought his corner doggedly. The politicking worked, and Ramanujan became a fellow of Trinity. Thus stimulated, he resumed work on mathematics.

But health remained poor, the English climate was suspect, and in April 1919, Ramanujan returned to India. The long voyage may have been a mistake, for by the time he arrived in Madras his health had once more deteriorated. On 26th April 1920 he died in Madras, leaving a widow but no children.

There are four main sources for Ramanujan's mathematics: his published papers, his notebooks, his quarterly reports to the University of Madras and his unpublished manuscripts. George Andrews of Pennsylvania State University found a fourth "lost" notebook – in actual fact a bundle of loose sheets – in 1976. Some manuscripts are still missing. Bruce Berndt, at the University of Illinois, Urbana, recently edited the first part of a three-volume work *Ramanujan's Notebooks*, aiming to supply proof of all his lost formulas.

Ramanujan had an unusual background, and no formal training. It was hardly a surprise that his mathematics was a little idiosyncratic. His greatest strength was in an unfashionable area – the production of ingenious and intricate formulas.

The formula states that, under suitable conditions on the function F, if

$$F(x) = \sum_{k=0}^{\infty} \frac{\phi(k)(-x)^k}{k!}$$

Then

$$\int_0^{\infty} x^{n-1}F(x)\,dx = \gamma(n)\phi(-n)$$

Here

$$F(n) = \int_0^{\infty} e^{-x}x^{n-1}\,dx$$

is Euler's gamma-function, which for integers n is equal to the factorial function $(n-1)! = (n-1)(n-2) \ldots 3.\,2.\,1.$

One curiosity of this formula is that $\phi(n)$ is defined only for positive integers n, yet the formula talks of $\phi(-n)$, Ramanujan is assuming that there is some "natural" extension of $\phi(n)$ that defines it for negative n – and in fact he often applies it when n is not a whole number.

His proof starts with Euler's integral representation of the gamma-function, expands the exponential function in a Taylor series, inverts the order of summation and integration, then inverts the order of summation, and finally applies Taylor's theorem again.

The conditions that Ramanujan states are insufficient to justify these procedures, so this proof of the master formula has gaps. To fill them, much stronger conditions must be imposed; but then the formula often cannot be used the way Ramanujan wishes to use it.

Despite these logical difficulties, Ramanujan nearly always gets correct results by using his master formula.

Figure 2.2

Ramanujan's master formula

Ramanujan was the "Formula Man" *par excellence*, unrivalled by any, save a few old masters: Leonhard Euler, Carl Gustav Jacob Jacobi. "There is always more in one of Ramanujan's formulae than meets the eye," said Hardy.

Most of Ramanujan's results are about infinite series, integrals, and continued fractions. He applied some of them to number theory. He had a special interest in analytical number theory, which seeks approximate formulas for such quantities as the number of primes below a given limit, or the average number of divisors of a given number.

His contact with Hardy influenced the work he published while at Cambridge, so is written in a relatively conventional style, with rigorous proofs. The results recorded in his notebooks have a very different quality. Because he was self-educated in mathematical research, his concept of proof was less than rigorous. If a mixture of numerical evidence and formal argument led to a plausible conclusion, and his intuition told him the answer was correct, that was enough for Ramanujan.

The advantage of no education

This is a common enough way for a professional mathematician to work out what the results probably are; but normally it is just the start of a lengthy and technical process of providing sound arguments to prove them. Ramanujan's methods leaped a few chasms, rather than being totally unorthodox. In fact his *results* were usually correct, but his proofs often had gaps. Sometimes a competent technician could fill the gaps; sometimes quite different arguments would be needed. Sometimes, but very seldom, his results were just wrong.

Bruce Berndt sums up Ramanujan's style like this:

"With a more conventional education, Ramanujan might not have depended on the original formal methods of which he was proud and rather protective ... If he had thought like a well-trained mathematician, he would not have recorded many formulas which he thought he had proved but which, in fact, he had not proved. Mathematics would be poorer today if history had followed such a course."

A good example is a result that Ramanujan called his "Master Formula" (see Figure 2.2). He was very fond of it and used it heavily. The master formula yields the value of the integral, between the limits zero and infinity, of $x^nF(x)$, for any function F. Ramanujan's proof involves various series expansions, changes in orders of summation, interchanges of the order of summation and integration, and so on. Because of the use of infinite processes, each step is fraught with danger. The greatest analysts spent most of the 19th century working out just when such procedures are permissible. The conditions that, according to Ramanujan, make his formula true, are grossly insufficient; and, as Hardy showed later, the formula holds only under much stronger hypotheses.

Consequently, on many occasions when Ramanujan makes use of his master formula, his arguments lack rigorous justification. Indeed, sometimes the master formula does not hold at all. Nevertheless, at the end of the day, almost all of the results that he gets are correct!

Some of Ramanujan's most striking work is in the theory of partitions, a branch of number theory. Given a whole number, we ask in how many ways it can be

The partition function $p(n)$ of n is the number of distinct ways in which n can be written as a sum of smaller (or equal) integers. For example, 5 boxes ☐☐☐☐☐ can be grouped as

☐☐☐☐☐	5
☐☐☐☐ + ☐	4 + 1
☐☐☐ + ☐☐	3 + 2
☐☐☐ + ☐ + ☐	3 + 1 + 1
☐☐ + ☐☐ + ☐	2 + 2 + 1
☐☐ + ☐ + ☐ + ☐	2 + 1 + 1 + 1
☐ + ☐ + ☐ + ☐ + ☐	1 + 1 + 1 + 1 + 1

giving a total of seven distinct partitions. So $p(5) = 7$. A short table of values of $p(n)$ runs as follows:

n	1	2	3	4	5	6	7	8	9	10
$p(n)$	1	2	3	5	7	11	15	22	30	42

At larger values of n the size of $p(n)$ grows rapidly:

n	50	100	150	200
$p(n)$	204 226	190 569 292	40 853 235 313	3 972 999 029 388

Figure 2.3

Partition functions

partitioned, that is, written as a sum of smaller whole numbers (see Figure 2.3). The *partition function* $p(n)$ is defined to be the number of ways to partition the given number n.

The numbers $p(n)$ grow rapidly with n. For instance $p(200)$ is a staggering 397 299 029 388. No simple formula for $p(n)$ exists. However, one can ask for an approximate formula, giving the general order of magnitude of $p(n)$.

A typical result in analytical number theory is the prime number theorem. This says that the number of primes less than x is approximately equal to $x/\log x$. The error is small in comparison, but becomes arbitrarily large as x becomes large. The approximation is good, however, in the sense that the ratio of the approximated value to the actual value tends to 1 as x tends to infinity: it is an *asymptotic* formula.

By its nature, the question of finding an asymptotic formula for the partition function is also a problem in analytical number theory. It is an especially intractable one for technical reasons, to do with how the partition function is defined. In 1918 Hardy and Ramanujan set about overcoming the technical difficulties, and obtained a formula.

Their result was a rather complicated series, involving 24th roots of unity. The first term of this series provides the initial approximation. When $n = 200$, this term yields the value 39 729 989 993 185.896, which agrees with the first six significant figures of the exact value. But they discovered that by adding on the next seven terms of their series, they obtained the value 397 299 029 388.004. As they say, this result "suggests very forcibly that it is possible to obtain a formula for $p(n)$, which not only exhibits its order of magnitude and structure, but may be used to calculate its *exact* value for any *n*". And they went on to prove precisely that. It must be one of the very few occasions when the search for an approximate formula has led to an exact one.

Ramanujan also found some remarkable patterns in the numbers of partitions. In 1919, he proved that for an integer k, the number $p(5k+4)$ is always divisible by 5, and $p(7k+5)$ is always divisible by 7. For example, let $k=2$. Then $5k+4=14$ and $7k+5=19$. We have $p(14)=135$, a multiple of 5; and $p(19)=490$, a multiple of 7.

In 1920, he stated some similar results: for example $p(11k+6)$ is always divisible by 11; $p(25k+24)$ is divisible by 25; all of $p(49k+19)$, $p(49k+33)$, $p(49k+40)$, and $p(49k+47)$ are divisible by 49; and $p(121k+116)$ is divisible by 121. Notice that $25=5^2$, $49=7^2$, and $121=11^2$. Ramanujan conjectured an even more general theorem, and said that as far as he could tell, formulas of this kind only exist for divisors of the form $5^a 7^b 11^c$.

This is a very strange set of results. There is nothing in the definition of $p(n)$, or the standard formulas for it, that suggest any special role for the three primes 5, 7, and 11. Proofs are, to say the least, elusive. Ramanujan didn't prove all of his conjectures, but those that he did prove led to the discovery of some beautiful combinatorial formulas. For example:

$$p(4)+p(9)x+p(14)x^2 + \ldots = \frac{5\{(1-x^5)(1-x^{10})(1-x^{15})\ldots\}^5}{\{(1-x)(1-x^2)(1-x^3)\ldots\}^6}$$

from which we can at once read off the result about $p(5k+4)$, for the right-hand side is manifestly a multiple of 5, whence each coefficient on the left must also be.

Some of Ramanujan's results remain unproved even to this day. One that succumbed during the past decade is particularly significant. In a paper of 1916, Ramanujan studied various arithmetical functions, in particular a function $\tau(n)$ defined to be the coefficient of x^{n-1} in the expansion of

$$\{(1-x)(1-x^2)(1-x^3)\ldots\}^{24}.$$

Thus $\tau(1)=1$, $\tau(2)=-24$, $\tau(3)=252$, and so on. I won't explain why this is a sensible thing to look at, except to say that it comes from deep and beautiful work in the 19th century on so-called elliptic functions.

Ramanujan was studying the following problem: *Let $\sigma_s(n)$ be the sum of the sth powers of the divisors of n. Find an approximate formula for the sum*

$$\sigma_r(0)\sigma_s(n)+\sigma_r(1)\sigma_s(n-1)+ \ldots + \sigma_r(n)\sigma_s(0).$$

He needed the function τ to express his results, and he also needed to know how big it was. He proved that its size is no larger than n^7, but conjectured that $n^{11/2}$ would suffice. He also conjectured various formulas, notably that

$\tau(mn)=\tau(m)\tau(n)$ if m and n have no common factor,

$\tau(p^{n+1})=\tau(p)\tau(p^n)-p^{11}\tau(p^{n-1})$ for prime p

Leo Mordel, the British mathematician proved these two formulas in 1919. They make it easy to compute $\tau(n)$ for any n. His ideas gave rise to some important areas of modern algebraic geometry. But Ramanujan's conjecture on the order of magnitude of $\tau(n)$ resisted all his efforts.

The innocent conjecture

In 1947, in Chicago, André Weil, an expert in number theory, was looking over some extremely old results of Carl Friedrich Gauss. He noticed how to apply them to study integer solutions of various equations. Following his nose, and a curious analogy with topology, he was led to formulate a detailed series of rather technical results, known as the *Weil conjectures*. Soon these acquired a central position in algebraic geometry. In 1974, the Belgian mathematician Pierre Deligne, in a *tour de force* of algebraic geometry, proved Weil's conjectures. A year later, he and Ihara applied the results to prove Ramanujan's conjecture too. It is a sign of the quality of Ramanujan's intuition, that his innocent-looking conjecture required such a massive and central breakthrough before it could be answered.

Because Ramanujan functioned in such an extraordinary manner, without formal training, getting correct results by unrigorous methods, people have suggested that there was something special about his thought patterns. Ramanujan himself is quoted as saying that the goddess Namagiri told them to him in dreams. However, he may have said this just to avoid embarrassing discussion. According to his widow, S. Janaki Ammal Ramanujan, he "never had time to go to the temple because he was constantly obsessed with mathematics". Hardy wrote that he believed "all mathematicians think, at bottom, in the same way, and Ramanujan was no exception". But he added: "He combined a power of generalisation, a feeling for form, and a capacity for rapid modification of his hypotheses, that were often really startling."

Ramanujan was not the greatest mathematician of his period, nor the most prolific; but his reputation does not just rest on his remarkable background and the touching "poor boy makes good" story. His ideas were influential during his lifetime, and they grow more influential as the years pass. Bruce Berndt believes that, far from being old-fashioned, Ramanujan was ahead of his time. It is sometimes easier to prove one of Ramanujan's remarkable formulas than to work out how he could possibly have thought of it. But by reconstructing where it came from, we can find new general principles. Only now, are we beginning to appreciate many of Ramanujan's deepest ideas. I leave the final word to Hardy:

> "One gift (that his mathematics) has which no one can deny: profound and invincible originality. He would probably have been a greater mathematician if he had been caught and tamed a little in his youth; he would have discovered more that was new, and that, no doubt, of greater importance. On the other hand he would have been less of a Ramanujan, and more of a European professor, and the loss might have been greater than the gain."

3 Saving lives

Sue Burns and Peter Wilder

This article gives a brief outline of the work of Florence Nightingale, and tells the story of a statistical investigation that she carried out during the 1870s into deaths among women giving birth.

Florence Nightingale is best remembered for her nursing in the military hospitals during the Crimean War between 1854 and 1856; but this was only a small part of what she achieved. Much of her work grew out of her passion for statistics as a way of convincing people of the need for action. In the Crimea, she initiated far-reaching improvements in medical hygiene. She also collected detailed information on every patient who passed through her hospital; careful analysis of this set of data enabled her to make a strong case for further improvements.

She was educated at home by her father, and she showed an early fascination with numbers. In 1845 she began working in hospitals, despite facing some opposition from her family, since hospitals were not considered to be fit places for ladies of good repute. In 1854 the War Office asked her to lead a party of nurses to the Crimea. On her return in 1856, she seemed to deliberately retreat from public life. The general public adored the 'lady with the lamp', but for Nightingale this adoration was an intrusion that would have made her real vocation impossible. She later said of the Crimea, 'I have seen hell and survived'. After her experience of being at the death of every one of thousands of men, Nightingale realised that not all these deaths were the inevitable result of wounds in battle. Many men could have recovered, but died from infection caused by unhygienic conditions. So she set out to ensure that such unnecessary deaths did not continue. In 1857, she reported to the Royal Sanitary Commission on her thirteen years' experience of visiting hospitals around Europe. The extent of her experience of hospitals was unparalleled at the time, and was all the more remarkable for being that of a woman in Victorian England. Her work as a dedicated statistician was recognised in 1858 when she was made a member of the Royal Statistical Society. Her most widely read book, *Notes on Hospitals*, was published in 1859 to help in the training of nurses, and many copies of this influential work still exist. In 1860 she opened her own school for nurses.

By this time her health was beginning to suffer, and she was almost entirely bed-ridden from 1860 onwards. In spite of her own poor health, she produced in 1862 a comprehensive report on the health of the British Army in India. This was an enormous undertaking for any one person, given the size of the British military commitment in India.

In the same year, 1862, Nightingale also opened a training hospital for midwives in London. Its purpose, in her own words, was 'to train midwives to work in hospitals and to deliver women in their own homes'. She noted that 'in nearly every country but our own there is a government school for midwives,' and she devoted considerable effort to this project in the face of opposition from the medical establishment. In the first five years 780 women gave birth, of whom 27 died. In 1867, there was a serious outbreak of puerperal fever, an infection of the uterus that was usually fatal, and nine women died within a few months. Eight of these deaths were due to the fever. Nightingale had no choice but to close the ward. Following this disappointment, she began to collect information about the deaths of women in childbirth from hospitals across the United Kingdom and Europe. At this time the causes of puerperal fever were unknown, and there were many competing theories. She began with the data shown in Table 3.1 from the first six years of her own training hospital in London.

Year	Deaths	Death rate	Number of deaths from puerperal fever
1862	3	1 in 32.3	3
1863	2	1 in 52.5	2
1864	3	1 in 47.0	1
1865	5	1 in 32.6	4
1866	5	1 in 30.0	2
1867	9	1 in 13.8	8

Table 3.1

Nightingale expressed the mortality rate in Table 3.1 as the average number of deliveries for one death. Later tables expressed the rate as the number of deaths per thousand deliveries. The mortality rate over the first four years in Nightingale's London training hospital was approximately 32 deaths per thousand deliveries. Florence Nightingale's first task in her investigation was to discover the rate of mortality in childbirth for the nation as a whole, to judge how extreme were the rates she had observed in her own hospital. By this time Nightingale was in regular correspondence with the leading statisticians of the day, including Sir William Farr, the Registrar-General. Farr gave her access to the following figures for England as a whole. The death rates in this table show the number of deaths per thousand deliveries.

From the data she received from Farr it was clear to Nightingale that the mortality among women giving birth in the first four years of her own training hospital was far greater than the average across the country; she was determined to find the reason for this. The information that she collected from other hospitals often did not include a separation by cause of death, but she was not happy to accept this without question. She tried to find ways of relating the data to the real situations from which they were drawn.

She was worried by the observation that, even before the outbreak of fever, mortality in her training hospital had been much higher than the rate for all births in England. One prominent politician of the time proposed that high death rates in hospital maternity wards were due to a high proportion of unmarried mothers giving birth in hospitals. It was commonly supposed that the mortality rate among unmarried mothers would be higher than among married mothers, because of the

All the data in Tables 3.2, 3.3 and 3.4 are drawn from Nightingale's report of her own investigation *Notes on Lying-In Institutions* (1872).

Total births	Deaths			
	Accidents in childbirth	Puerperal disease	Chest diseases	TOTAL deaths
768 349	2346	1203	154	3933
Death rate in 1867	3.10	1.60	0.2	5.10
Death rate for 1855-67	3.22	1.61		4.83

Table 3.2

Mortality after childbirth in England, 1867 (Registrar-General's Annual Report)

perceived immorality involved. This hypothesis was thoroughly refuted by information that Nightingale collected from workhouses in London and Liverpool. One Liverpool workhouse recorded 6396 births in 13 years, of which 936 were to unmarried mothers; but there were only 6 deaths, a mortality rate of only 4.2 per thousand. Nightingale summarised her data from hospitals and workhouses, and compared them with national figures over an equivalent period, shown in Table 3.3.

Unmarried mothers were frequently sent to workhouses. Therefore the proportion of unmarried mothers giving birth in workhouses was greater than in the population as a whole, or in hospital births. So, if being an unmarried mother made death in childbirth more likely than normal, the data would show a greater number of maternal deaths in workhouses.

But Nightingale's table shows the opposite. The death rate in workhouses is close to that for 'all England', in both Tables 3.2 and 3.3. And the maternal death rate in hospitals is much greater than for the workhouses.

What else did Nightingale notice from her data? Hospital maternal death rates significantly exceeded the national average and the workhouse rates. Something was affecting hospitals but not workhouses. But the difference was most striking for puerperal disease; for other causes of death, the rate was not much different between hospitals, workhouses and other places.

The term 'lying-in hospital' refers to a maternity hospital.

This suggested that in hospitals, the most common cause of death in childbirth was puerperal disease.

Places	Puerperal disease	Accidents in childbirth	TOTAL
King's College Hospital lying-in ward	29.4	0	29.4
12 Paris hospitals: 1861	-	-	75.2
1862	-	-	56.7
1863	-	-	60.6
Queen Charlotte's lying-in hospital (40 years)	14.3	5.3	19.6
27 London workhouses in which both delivery and death occurred	4.1	2.1	6.2
40 London workhouses, including those without deaths (5 years)	3.3	1.7	5.0
Liverpool workhouse lying-in wards (13 years)	3.4	2.2	5.6
All England (13 years)	1.61	3.22	4.83
All England: 64 health districts (312 402 births in 10 years)	-	-	4.3
All England: 11 large towns (10 years)	-	-	4.9
8 Military lying-in hospitals (2-12 years)	3.9	3.4	7.3

Table 3.3

Mortality per 1000 births, from accident in childbirth and puerperal disease

Nightingale went on to compare the data in Table 3.3 with similar data collected by a French doctor about death rates among women who gave birth in their own homes. The figures in Table 3.4 show that the mortality rate amongst home births was very close to the average rate of mortality in childbirth for the population as a whole. Nightingale was therefore able to conclude that the risk of contracting puerperal disease was very much greater in hospital than anywhere else.

The figures in Table 3.4 are taken from M. le Fort's Tables in *Des Materaites*

Places	Years observed	Deliveries	Deaths	Death rate
Edinburgh	1	5186	28	5
London:				
Westminster general dispensary	11	7717	17	2
Westminster Ben Institute	7	4761	8	1
Royal Maternity Charity	5	17 242	53	3
London population	5	562 623	2222	3.9
St Thomas' Hospital	7	3512	9	2.5
Guy's Hospital	8	11 928	36	3
Guy's Hospital	1	1505	4	2
Guy's Hospital	1	1702	3	1.7
Guy's Hospital	1	1576	11	6
Paris:				
12th Arrondissement	1	3222	10	3
Bureau de Bienfaisance	1	6212	32	5
Bureau de Bienfaisance	1	6422	39	6
City of Paris	1	44 481	262	5
City of Paris	1	42 796	226	5
Leipzig Polyclinique	11	1203	13	10
Berlin Polyclinique	1	500	5	14
Munich Polyclinique	5	1911	16	8
Greifswald Polyclinique	4	295	6	20
Stettin Polyclinique	17	375	0	0
St Petersburg Polyclinique	15	209 612	1403	6.6

Table 3.4

Death rates from all causes: Women delivered in their own homes

At the end of her investigation Nightingale drew the following conclusion:

> 'In estimating the probable accuracy of statistical data in which there may be both excesses and deficiencies, sources of error are diminished by largeness in the numbers employed in striking averages. Bearing this in mind, and after considering the objections brought against the accuracy of the figures, there seems no reason for rejecting the Registrar-General's average total mortality among lying-in women in England of 5.1 per 1000, as affording a sufficiently close approximation to the present real death rate among lying-in women delivered at home, for all practical purposes of comparison with death rates in lying-in hospitals.'

In this quotation Nightingale appealed to the Law of Large Numbers to support her reliance on a death rate of 5.1 per 1000 births as a good estimate of the underlying rate of maternal mortality in England. In particular, all her data pointed to the conclusion that the chance of a mother dying in childbirth was much higher if the birth took place in a hospital maternity ward than if the delivery was at home. Nightingale concluded, 'For every 2 women who die at home, 15 will die in hospital.'

This conclusion drew Nightingale to ask herself why the mortality rate among women giving birth in hospitals, and especially training hospitals, was so much higher than among women giving birth at home, or in workhouses. She collected information from hospitals and workhouses about practices that she thought might affect the spread of disease. One hospital, which had an unusually low death rate, was asked to describe its practice when a case of fever was discovered. The feverish case was transferred to an isolation room. The ward was closed and then 'fumigated, cleansed and lime washed' before being re-opened. Nightingale discovered that it was common practice in the workhouse, unlike the hospitals, for the cells to be thoroughly scrubbed and cleansed before the next inmate was admitted. She concluded that this practice prevented the possibility of infection being passed on to other inmates. When a woman gave birth at home, she would be the only woman to use the bed so there was no possibility of passing infection on to another.

More remarkably, Nightingale discovered that, in training hospitals, almost the only people who freely moved from one part of the hospital to another were the students, who were expected to see all aspects of the life of the hospital. In some circumstances, the mortuary was next door to the labour ward, and the student nurses were the only people who regularly went straight from one to the other. She concluded that the students might be responsible for the transmission of disease in the training hospitals, and that they might therefore be contributing to the extremely high rate of mortality in institutions such as the Paris hospitals, which all had student doctors and nurses attached to them.

Nightingale presented the findings of her investigation in a short book called *Notes on Lying-In Institutions* in 1872. In this book she gave a summary of her main conclusions, from which the following are drawn.

- The death rate in lying-in wards is many times that for home delivery (allowing for inaccuracies).
- The greatest cause of death in hospital lying-in wards is 'blood poisoning'. This implies that in these wards there are conditions and circumstances that aid the spread of infection.
- The risk of puerperal disease is increased by:
     crowding many cases into one room;
     keeping them too long in the same room;
     using the same room too long without cleaning, evacuation, rest, or airing.
- Other factors contributing to the prevalence of puerperal disease in some hospitals are:
     prevalence of puerperal fever outside;
     midwifery wards within general hospitals;
     proximity of post-mortem theatres or other external sources of putrescence;
     admitting medical students from general and anatomical medical schools;
     treating cases of puerperal disease alongside midwife cases;
     using same attendants, bedding and nurses in infirmary wards/lying-in wards;
     lack of ventilation, cleanliness and attendants.

Nightingale made a number of substantial recommendations for the reform of midwifery and for the design of labour wards. Far from recommending that all births should take place in the mother's own home, she prepared detailed drawings of how a 'Lying-in Hospital' should be designed. All the rooms were to be 'light and airy', with plenty of space around the beds, and none of the rooms was to contain more than four beds. She also recommended that information should be recorded after every birth, and she designed a detailed form for this purpose. This form was to be completed by the midwife, to provide detailed information about the mother, the period of labour, and the infant. It even included the following requirement: 'Should any death occur within a month, the date and cause of death are to be entered, together with any abnormal configuration or conditions of health which may influence the result of the delivery.' Nightingale had proposed a similar idea in 1860, when she developed a Model Hospital Statistical Form, with help from prominent doctors. Unfortunately, the Hospital Form of 1860 proved to be too complicated at the time, and the scheme was never put into general use.

Although her ideas were not immediately adopted, they have since become the foundation for modern medical statistics. Uniform and accurate hospital statistics, she wrote, would 'enable the value of particular methods of treatment and of special operations to be brought to statistical proof.' She saw the need to collect information, not only about the immediate outcome of the labour, but also about the longer term. She recognised the potential for improving the quality and effectiveness of health care by collecting and analysing detailed and carefully structured information. In the case of a death, she saw it as essential to identify and record the cause of death clearly. When she collected data, she was able to present it to the influential people of her day to make a powerful case for the improvement of living conditions. Her enthusiasm for statistics was passionate, and she was dedicated to using her passion to identify the causes of 'human misery and suffering', and to enable change.

In this approach to statistics, Nightingale was quite different from the Registrar-General. He was typical of the attitude of the time. He wrote to her in 1861:

> 'We do not want impressions, we want facts. Again I must repeat my objections to intermingling Causation with Statistics ... The statistician has nothing to do with causation; he is almost certain in the present state of knowledge to err ... You complain that your report would be dry. The drier the better. Statistics should be the driest of all reading. What I complain of is that on reading your report I am conscious of receiving a wrong impression, because your details are not sufficiently dry and sufficiently extensive.'

This emphasis on dry facts, at the expense of impressions and arguments about causation, may have arisen from fears of political revolution, which were current at the time. Nightingale's approach to statistics was in complete contrast to this. She wrote later to the Belgian statistician, Quetelet, about why her study of practical applications of statistics mattered so much to her.

> 'On my part this passionate study is not based on a love of science, a love I would not pretend I possessed. It comes uniquely from the fact that I have seen so much of the misery and sufferings of humanity, of the irrelevance of laws and of Governments, of the stupidity, dare I say it? – of our political system, of the dark blindness of those who involve themselves in guiding our body social that ... frequently it comes to me as a flash of light across my spirit that the only study worthy of that name is that of which you have so firmly put forward the principles.'

4 Tartaglia – 'The Stammerer'

David Tall

Dirty work afoot in cubic equations

You might read this article about cubic equations when you learn about the quadratic equation formula. You might also enjoy reading about the intrigue and professional rivalry between Tartaglia and Cardano.

This method of solution is relevant to the history of the development of complex numbers. You may find it interesting to compare this mathematics with that of the final section of Chapter 1 in the *Complex numbers and numerical methods* option or with Omar Khayyam's work in the *History of mathematics* option.

In the lengthy history of mathematics the work of Niccolo Fontana (1499–1557) occupies just one brief moment of fame, but it is a tale of intrigue and double dealing which would tax the little grey cells of Agatha Christie's Hercule Poirot. The story begins in the early 16th century when the twelve-year-old Niccolo was accidentally caught up in an attack by the French on his home town of Brescia, Italy. He was severely injured by a savage sword-thrust in the mouth and it was only the careful nursing by his mother that saved him from death. The incident left him scarred not only through physical deformity, but also through damage to his speech resulting in his nickname, Tartaglia – 'The Stammerer'.

Born into a poor family (his father was the local postman in the village) he had an intelligent and enquiring mind and was able to teach himself the rudiments of mathematics and physics. In 1516 he moved to Verona, at first to teach the abacus, and then from 1529–1533 as principal of a school. He later moved to Venice and became a Professor of Mathematics.

At the time there was a blossoming interest in the solution of equations. The quadratic equation had been solved, giving the usual formula which is now taught in the secondary school National Curriculum, so the knowledge of algebra at that time was at, or about, the level of the modern GCSE syllabus. But there was a difference. First, what we now regard as the arithmetic of numbers was then regarded as the geometry of lengths, so that a quadratic equation involved areas and cubic equations concerned volumes. Second, instead of using brief algebraic notation, the prevalent symbolism of the day was but a shortened form of everyday language. What we might write as

$$x^3 + 6x = 20$$

was then written as

> *cub p: 6 reb aeqlis 20*

where *cub* stood for the cube of the unknown, *p*: (piu) for plus and *reb* for the unknown itself. As numbers were lengths, negative numbers were not recognised as having any meaning (how can you have a length less than nothing?), so general equations were always written in such a way to have all plus signs on one side of the equation or the other. What we might write as

$$x^2 - 2x + 3 = 0$$

would be recast in the form of

$$x^2 + 3 = 2x$$

so that even the familiar quadratic would be considered as several different types of equation, depending on which side the various terms occurred.

By the time Tartaglia was living in Venice the search was on for the solution of the cubic equation, or rather, of the various kinds of cubic equation, such as (in our notation)

$$x^3 + ax = b$$
$$x^3 = ax + b$$
$$x^3 + ax^2 = bx + c$$
etc...

Some of these forms are easier to solve than others and so a variety of methods was developed by various individuals and there were contests in which mathematicians challenged each other to mathematical duels. Each set the other problems and the winner was the one who solved the most.

The scene was set for Tartaglia, who was not one for verbal dispute. The opportunity to test his wits against others through the giving and solving of written problems gave him an ideal public stage. In one such contest against Antonia Maria Florido of Venice, Tartaglia emerged the victor by solving all the problems set before him whilst Florido could solve none.

For a brief time Tartaglia was a celebrity – a magician who could do things that baffled others. His secret lay in his ability to solve equations of the type $x^3 + ax^2 = b, x^3 = ax + b$ and $x^3 + b = ax$. Although Florido knew how to solve the first, he could not cope with the other two.

This power over others could not last. Mathematics is not like that. Today the rule is 'publish or perish' and the great men are those who give their new found knowledge to the world. But in those days the power to do something that others could not was highly prized. Another mathematician, Cardano, begged Tartaglia to reveal his secret, and Tartaglia at first refused. Then vanity got the better of him. He would not give a straight answer but, after obtaining a pledge of secrecy from Cardano, he wrote an obscure verse hinting at the solution of $x^3 + ax = b$:

> Quando che'l cubo con le cose appresso
> Se agguaglia a qualche numero discreto
> Trovan dui altri, differenti in esso...

(When x^3 together with ax are equal to a number b, take two other numbers, essentially different and ...)

The hint was enough – Cardano (physician, mathematician, scientist-in-general and inveterate gambler) used his wits to crack the cubic code. He reneged on his pledge of secrecy and published his complete set of solutions of the cubic equation in 1545 as *Ars Magna*. It is a book which remains in print today for its historical significance. It includes 13 chapters devoted to different forms of cubic equation.

For example, Cardano notes that to solve

$$x^3 + ax = b$$

in modern notation we can make progress by seeking u, v such that

$$3uv = a \text{ and } u^3 - v^3 = b$$

for then, if $x = u - v$, we have

$$
\begin{aligned}
x^3 + ax &= (u - v)^3 + 3uv(u - v) \\
&= u^3 - 3u^2v + 3uv^2 - v^3 + 3u^2v - 3uv^2 \\
&= u^3 - v^3 \\
&= b
\end{aligned}
$$

so $x = u - v$ is a solution of the original cubic.

To solve, for example, $x^3 = 6x = 20$, we require u, v where

$$3uv = 6, \; u^3 - v^3 = 20$$

so we can substitute $v = 2/u$ from the first equation into the second to get

$$u^3 - \frac{24}{u^3} = 20$$

and, for $k = u^3$, this gives

$$k - \frac{24}{k} = 20$$

which simplifies to a *quadratic* in k:

$$k^2 - 20k - 24 = 0$$

with solutions

$$k = \frac{20 \pm \sqrt{\left(20^2 + 4 \times 24\right)}}{2}$$

or

$$k = 10 \pm \sqrt{108}$$

Thus

$$u^3 = 10 \pm \sqrt{108}$$

and then

$$v^3 = u^3 - 20 \text{ so } v^3 = \left(-10 \pm \sqrt{108}\right)$$

One root can therefore be found in the form

$$
\begin{aligned}
x &= u - v \\
&= \sqrt[3]{\left(10 + \sqrt{108}\right)} + \sqrt[3]{\left(-10 + \sqrt{108}\right)}
\end{aligned}
$$

Once one root of a cubic is found, it gives a linear factor and the remaining quadratic part is easy.

By such an ingenious route, Tartaglia discovered his solution and his clue to Cardano was sufficient for his contemporary to steal his secret and write the first widely acclaimed algebraic text-book. It remained pre-eminent well into the next century until Descartes (inventor of cartesian coordinates) linked algebra and graphs in the way that we know today. The book gave Cardano fame and fortune but little credit was given to the man who made the vital breakthrough. It is the way of academic life. The honour goes to the person who publishes first.

Tartgalia went on to further conquests. He made the first translation of Euclid into a modern language (Italian) in 1543 and he proved the theorem that the maximum range of a projectile is achieved by a firing elevation of 45°. Yet he failed to make a tangible profit from his skills and died alone, in poverty, near the Rialto Bridge in Venice on 15 December 1557.

5 Geometry, shoemaking and the Milk Tray problem

Ray Cuninghame-Green

An important idea behind the problems in this article is that in order to solve them using a computer you have to design an algorithm. The methods used here are virtually the same as those used in the *Algorithms* unit in Book 1. The difference here is that you also have the added problem of representing the geometric shapes by numbers so that you can manipulate them, just as you do in the unit on space and vectors.

Cutting out garments from material, printing wallpaper, filling a chocolate box or programming a robot all involve a tricky task – moving and fitting together irregular shapes econo-mically. Mathematicians are designing computer programs to help manufac-turers to plan for profit.

When I moved to Birmingham where I now live, three men packed the contents of my house into a vehicle small enough to be allowed on the roads. At my new home, a carpenter was cutting sheets of veneered chipboard to make cupboards for the kitchen, when my piano arrived and had to be manoeuvred round corners and though doorways without gouging out lumps of plaster from the walls.

Fitting the cupboards, filling the van and shifting the piano pose geometrical problems of cutting, packing and motion planning. Although they seem different, they are closely related: they are all concerned with finding arrangements, or motions, of shapes within a confined space. The shapes must not violate one another's share of the space by colliding or overlapping, nor penetrate the boundaries of the space.

The best way of fitting shapes together merits serious study, not just for moving house, but also because it is commercially important to industry. Shoemakers, for example, try to cut out pieces of leather from a single skin to make the most economical use of the material. Figure 5.1 shows copies of a horseshoe-shaped component for shoes, known as a vamp, cut from sheets of material. If the grain of the material requires all the shapes to lie with the same orientation, then you need the layout such as that shown in Figure 5.1a. Here, about 46 per cent of the material goes to waste. If the grain allows shapes to be turned upside down, a layout such as in Figure 5.1b reduced the wastage to 33 per cent. The interlocking layout of Figure 5.1c is even better, with only 20 per cent of the material wasted. Similar cutting problems arise in many different industries: think of the manufacture of clothing, or of stamping tin lids out of sheet metal.

Packaging objects efficiently is also extremely important in commerce; for instance, chocolates with assorted shapes must be arranged economically on a tray, or crates of different sizes in a warehouse must be stacked in a tidy, stable pile. Again, the problem of motion planning arises on a modern assembly line in instructing robots to manipulate components without collision.

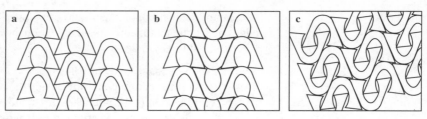

Figure 5.1

Fitting shapes together economically is important in industry

These problems are rather messy and complicated. The objects involved have strange shapes, the space they occupy or move in may be three-dimensional with complicated boundaries, and the movements that can be made are difficult to classify. Nevertheless, by considering a drastically simplified arrangement we can see how geometry can provide a way of dealing with such problems. We can then derive a few general principles and use a computer program to extend them to more complicated situations. Figure 5.2a shows a quadrilateral Q, fixed on the page, and a triangle T, free to slide about on the page but without rotating. One point in the triangle T is labelled as a point of reference – the "datum" point P.

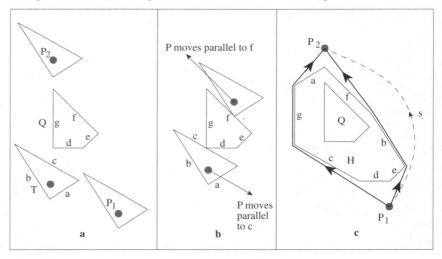

Figure 5.2

To pose a cutting problem, suppose the page is a sheet of material from which we wish to cut a copy of the quadrilateral Q and a copy of the triangle T in orientations shown in Figure 5.2. How can we describe all the possible positions for T such that T and Q do not overlap? Or suppose we wish to slide triangle T from position P_1 to position P_2. During this process, the triangle may touch, but must not penetrate, quadrilateral Q. Can we characterise the permissible motions of T?

To gain some insight, imagine we take triangle T and move it without rotating to some position where it just touches quadrilateral Q; then "wipe" T around Q. That is to say, we move T round Q without rotation, constantly remaining in contact with Q but not breaking its boundaries. When a vertex of T slides along an edge of Q, the datum point P moves parallel to that edge; but when an edge of T slides on a vertex of Q, the datum point moves parallel to the edge of T (see Figure 5.2b). So the path of the datum point consists of copies of the four edges of Q and the three edges of T coupled together in suitable sequence, giving the heptagon, debfagc, labelled H in Figure 5.2c.

You can soon see that if we place T without rotating it anywhere on the page, then T overlaps Q if the datum point falls inside the heptagon H; T touches Q if the datum point falls along the boundary of H; and T and Q do not meet if the datum point falls outside H. The construction of H thus enables us to show the permissible positions for T in the cutting problem. We can now reformulate the motion-planning problem in the following terms: find a path, joining the initial and final positions P_1 and P_2 of the datum point, not crossing the interior of H. In Figure 5.2c, path s is clearly one possibility, as are the two paths marked [with arrows], passing one either side of H.

The heptagon H is called the configuration space obstacle, or CSO, for the triangle T relative to the quadrilateral Q. We can easily translate the physical process of wiping T round Q into a set of rules for calculating the CSO, given descriptions of T and Q (see Calculating the configuration space obstacle on page 28). These rules work not just for a quadrilateral and a triangle, but also for any pair of convex polygons: that is to say, straight-sided figures where all the vertices point outwards – unlike the nonconvex hexagon shown in Figure 5.3a, for example.

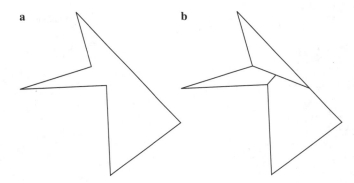

a   b

Figure 5.3

*This figure shows a nonconvex hexagon **a** with a two vertices that point inwards. To calculate the CSO, we dissect the hexagon in **b** into three convex components*

What about curved shapes? The [striped] shape resembling a circle in Figure 5.4 is actually a 20-sided polygon, but because this 20-gon approximates to a circle, it is not too difficult to accept that we can extend the rules to calculate CSOs when circles or other convex shapes with smoothly curving boundaries are involved, though the method now needs to be expressed using differential calculus. For most practical purposes, however, we can just replace any such object by a polygon that is almost identical.

In Figure 5.4, these ideas are shown in a computer program called FITZ, which can deal with a version of the chocolate-tray problem mentioned before. The user defines a set of shapes and also stipulates the width of a tray. The program first calculates the CSO for each pair of shapes, to avoid them overlapping; it then packs the shapes, without rotation or overlap, onto the tray, trying to use the least length.

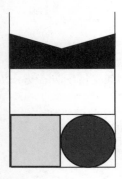

Figure 5.4

To deal with the chocolate tray problem, we turn curved shapes into polygons that are almost identical

Suppose instead that we have a roll of material of given width from which we want to cut out the shapes of Figure 5.4, in the given orientations, using the material as economically as possible. Obviously, Figure 5.4 will do as a cutting pattern. So in fact, there is no essential difference between a cutting problem and a packing problem.

Unfortunately, the shapes occurring in many practical problems are often not convex. They may have vertices pointing inwards as in Figure 5.3a. One way out of this difficulty is to enclose each nonconvex shape within a larger, convex shape. Any arrangement of the new convex shapes that avoids overlap is obviously allowed for the original nonconvex shapes. We must be cautious, however. If the convex shape within which we embed our nonconvex shape is unnecessarily large, we shall lose too much of the character of the original problem. We need the smallest convex figure within which we can embed our original shape.

There is a neat way of thinking about this problem. Consider the cogwheel of Figure 5.5. If we slip an elastic band under tension around the cogwheel, the elastic spans across the dents in the cogwheel, creating a convex shape, which we can call a cam (see Figure 5.5c). It is not too hard to show that no smaller convex shape can contain the cogwheel; we say that the cam is the "convex hull" of the cogwheel. Again, a physical process with an elastic band is not the same thing as a calculation, but, in fact, the calculation of convex hulls of arbitrary shapes is a problem for which mathematicians have discovered very efficient methods.

Replacing shapes by their convex hulls give us, then, a possible approach to cutting and packaging problems for nonconvex shapes. As a bonus, it is also relevant to the motion-planning problem in Figure 5.2, where we seek paths joining P_1 to P_2, not crossing the interior of the CSO of the heptagon H. In Figure 5.2c, we imagine pins driven into the page at P_1, P_2, and at each vertex of H; then an elastic band enclosing these nine pins, under tension. The elastic band will assume the form of the convex hull, simultaneously defining two alternative paths from P_1 to P_2 avoiding the interior of H.

In an industrial situation, we should opt for paths that are short, thereby economising on time and work. If we consider some path, say s in Figure 5.2c, and think of it as a loose string, we can steadily pull the string tighter to produce shorter and shorter paths until the string becomes as tight as the elastic band. Thus the two paths defined by the convex hull are as short as possible among those that "keep right" and those that "keep left" respectively round H (see Figure 5.6).

Unfortunately, important as the convex hull is, it does not answer all our needs. If we replaced the shoe vamps in Figure 5.1 with their convex hulls, we should lose the possibility of interlocking them and so waste material. A better approach is to visualise nonconvex figures as divided into smaller convex figures – rather than embed them in larger ones.

Packing very irregular shapes

Thus, to calculate the CSO of our triangle T relative, say, to the nonconvex hexagon of Figure 5.3a, we first dissect the hexagon into convex components as shown in Figure 5.3b, and calculate separately the CSO of the triangle relative to each component. We can then calculate the required CSO from the component CSOs. And if, instead of the triangle T, we had been given some nonconvex figure,

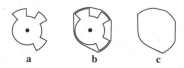

Figure 5.5

*For a nonconvex shape, such as a cogwheel **a**, we need to find the smallest convex shape, that fits around it. An imaginary elastic band in **b** spanning the dents gives the "convex hull" in **c**, which is cam–shaped*

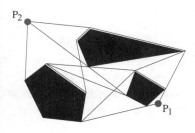

Figure 5.6

A network of the possible paths, avoiding several obstacles, for the situation in Figure 5.2

we should have dissected this too, into convex components and related each of these to each of the convex components of the nonconvex hexagon.

In this general way, we may arrive at a technique for cutting and packing highly irregular, non-convex shapes. For example, in Figure 5.7a, five irregular shapes must be cut frm a board, 12 units wide. Figure 5.7b shows how the computer program FITZ arranges them so as to use only an 11-unit run of board.

It is not hard to find a systematic method – an algorithm – for dissecting the polygon into convex shapes. We simply apply the following procedure in turn to each pointing-inwards vertex of a nonconvex polygon: "Bisect the interior reflex angle and extend the bisector until it meets a line already present".

Figure 5.3b shows this procedure used to dissect the nonconvex hexagon into three convex components. If the original figure has v vertices pointing inwards, this algorithm produces $(v+1)$ components. Because each component has to be used in order to calculate CSOs, the question is whether this simple algorithm produces a dissection with the least number of components. The answer is no, but this is not the whole story. Can we then find an efficient way of dissecting a nonconvex polygon into the least number of convex components?

Here we enter deep waters: such algorithms tend to be very intricate, and practical computer programs for industrial application will usually avoid the full rigours of this kind of mathematical analysis. Nevertheless, we can write programs for many applications involving complicated shapes, such as designing wallpapers. A wallpaper pattern is a regularly repeating two-dimensional arrangement of a motif. The cutting layouts of Figure 5.1 are essentially wallpaper patterns. There is a well-known proposition called Fedorov's theorem. It tells us that there are only 17 different types of wallpaper pattern. The Hungarian mathematician L. Fejes Tóth invented mathematical labels for them based on the letter W. A pattern, such as the one in Figure 5.1a, is of type W_1.

The reason that such regular layouts are important in shoe manufacturing is that the cutting is carried out by a machine that is relatively limited in how it moves. A cutting head travels across the material punching out a row of copies of the shape, then travels back punching out another row. On some machines, the cutting head can rotate through 180° and follow a cutting layout like those of Figure 5.1b and 5.1c, which are of type W_2 in Fejes Tóth's notation.

If the shapes are convex and cut from large sheets of material, then Fejes Tóth showed that for a class of shapes the best W_1 arrangement will never use the material more efficiently than the best W_2 arrangement. In the practical world of nonconvex shapes and finite sheets of material, this is still frequently, though not always, the case.

Finally, now that we have seen how to deal with nonconvex figures, we can return to the problem of practical motion planning; there is obviously a considerable gulf between an artificial problem of moving a triangle around a quadrilateral in a plane and a practical problem, such as devising a sequence of motions whereby an industrial robot can manipulate a complex object into position on an assembly line. How do the artificial and the practical problems differ?

First, there will usually be several obstacles to be avoided, rather than just one. Figure 5.6 shows a network of strings connecting the points P_1 and P_2. Each string

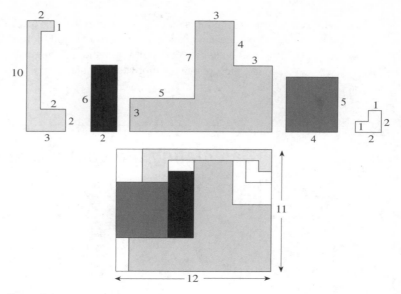

Figure 5.7

Cutting shapes economically from a board

corresponds to a way of passing from P_1 and P_2 keeping to the left of certain obstacles and to the right of others, and they are pulled tight so as to be as short as possible. Finding the shortest route from P_1 to P_2 now requires finding a path through a network. But this task is already well understood by mathematicians who have devised efficient algorithms.

The artificial problem also ignores the possibility that there may be given boundaries of the space, but this presents no difficulty. Such boundaries can simply be thought of as the edges of very large obstacles, giving no new problems of principle. We have also seen how to deal with curved or nonconvex boundaries.

Undoubtedly, the greatest limitation of the artificial problem is that it deals only with very special motions: translations in the plane. Most of what we have discussed can, however, be generalised. For example, we can define convex hulls for three-dimensional objects and calculate them efficiently. But we must go further than this. When we move the triangle T about the plane, without rotation, the positions taken by T can be specified precisely by giving the coordinates (x, y) of the datum point. The collection of all the possible arrangements of T relative to Q is called the configuration space. Thus, for this problem, the configuration space is essentially the same as the physical space – the page within which the action takes place: both are describable by a collection of number-pairs (x, y).

Suppose now that we allow the triangle to rotate in a certain way during the motion. To describe the position of the triangle at any time now requires three numbers (x, y, q) where q is its angular orientation. Now the configuration space is three-dimensional even though the physical space is still only two-dimensional.

In motion planning for a robot, the physical space is three-dimensional but because rotations as well as translations are involved, the configuration space is actually six-dimensional. Like all hard problems, this has stimulated more than one plan of attack and it is still the focus of much research. Identifying the fundamental geometrical ideas, such as those behind motion planning, is one of the ways mathematicians are helping manufacturing industry to become more efficient and use their resources more economically.

Calculating the configuration space obstacle

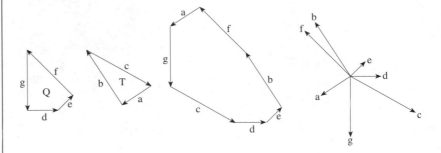

Make the edges of the quadrilateral Q and the triangle T into arrows, such that if you walk round the boundary of Q following the arrows, the interior of Q is always on your left, whereas if you walk round the boundary of T following the arrows, the interior of T is always on your right.

Bundle together the seven arrows so created, by their blunt ends at any point O, keeping each arrow pointing in its original direction. In this example, the arrows lie, in anticlockwise order round O, in the sequence debfagc. Now connect the arrows end-to-end in the same sequence debfagc, still without changing direction. The result is the CSO of T relative to Q.

The method works for any pair of convex polygons. Actually, as described, the method just gives the shape of the CSO, but a slight elaboration suffices to calculate its position as well.

6 A tale of two cheeses

Adrian Oldknow

Adrian Oldknow is Head of
Mathematics at the
Mathematics Centre, West
Sussex Institute of Higher
Education.

This article presents a series of problems about the growth of bacteria on a falling cheese. The method of solution is the same as the Euler step-by-step method, which you can find in the *Reversing differentiation* unit in Book 2, although it is presented in a different way in the text of the unit.

It was a lovely summer's day. The hot-air balloon seemed to hang stationary way above the open fields. The picnic was laid out on a spotted cloth on top of the fuel tank in the basket. There was fruit, wine, bread and cheese. In fact there were two large chunks of cheese. One was a rich, creamy round of Camembert, the other was a fine piece of strong Cheddar. Just for a moment the balloon shuddered as it entered an air pocket. The captain rushed to light the burner. At the moment she brushed against the Cheddar cheese which fell over the edge of the basket, while the heat of the burner awoke the dormant Listeria bacteria which had been sleeping quietly in the Camembert. What happened next?

Chapter 1: The flight of the Cheddar – (expurgated version)

The acceleration of the Cheddar is that of gravity – i.e. $g = 9.8$ m/s/s approximately (as observed by Galileo one sunny afternoon in Pisa). Assuming that the balloon was stationary and that the cheese left the balloon at time $t = 0$ with downward velocity $v = 0$ then make up a table of its velocity v m/s after times $t = 1, 2, 3, …, s$.

t	0	1	2	3	4	5
v	0	9.8				

What would a graph of v against t look like? If the speed of sound in air (Mach 1) is about 330 m/s how long will it be before the Cheddar breaks the sound barrier? Is this likely? Give some reasons …

Chapter 2 : The growth of the Listeria – (expurgated version)

Scientists have shown that Listeria bacteria in unpasteurised cheeses at about 25°C increase at a rate of about 20% per second. The average number of dormant Listeria in such a piece of Camembert would be 1000. Assuming that this is the count $c = 1000$ when the Cheddar started to fall $\left(\text{at } t = 0\right)$ then make up a table of the count c after times $t = 1, 2, 3, …, s$.

t	0	1	2	3	4	5
c	1000	1200				

What would the graph c against t look like? How many Listeria bacteria would be on the Camembert at the time the Cheddar broke the sound barrier? Is this likely? Give some reasons …

Chapter 3: Meanwhile back on the Cheddar ...

We have seen how the velocity v of the Cheddar could change with time, but how about the distance d through which it falls? This is a bit trickier since velocity is a continuously changing quantity (but isn't that also true of the increase of the bacteria?). However, we could reckon that the distance fallen in any second is given by its average velocity in that second (why?) and so build up a new table:

t	0		1		2	3	4	5
v	0		9.8		19.6			
av		4.9		14.7				
d	0		4.9		19.6			

So how far (in metres) would the Cheddar have fallen before it reached Mach 1? What does a graph of d against t look like? What shape is it? Can you find formulae for the velocity v and the distance d after time t?

Chapter 4: Limits to growth: the Listeria

Chapter 2 used a model of population growth often attributed to the Rev. Thomas Malthus, an eighteenth century English cleric. This does not take into account the finite size of the Camembert in question and its incapacity to sustain a population of bacteria greater than 10 000. The modified model (due to Verhulst) assumes that the rate of growth decreases (or is damped) as the count c gets closer to the maximum $m = 10\,000$ by a factor $(1 - c/m)$. If the rate of growth without resource constraints is r (as a decimal) then the formula for the rate of increase at a time t when the count is c is given by: $r.c.(1 - c/m)$. If we assume, as before, that the increase is roughly constant for our little intervals of 1 second then we can build a modified table:

t	0	1	2	3	4	5
c	1000	1180				

How many bacteria will there now be at the time the Cheddar breaks the sound barrier? How long does it take the count to reach 99% of its maximum? What does the graph of count c against time t now look like? This is known as a **logistic curve**.

Chapter 5: Limits to growth: the falling Cheddar

Chapter 1 used a model of acceleration which might be appropriate in a vacuum. However Cheddar is not very aerodynamic (drag coefficient = ??), and the nice warm summer's air would resist the cheese's motion, giving a deceleration which reduces the acceleration due to gravity. This resistance increases with the speed of the cheese and acts in the opposite direction to its velocity. At some critical speed the deceleration due to air resistance will cancel out the acceleration due to gravitational attraction entirely and the Cheddar will then have a constant velocity (called the terminal velocity). The refined model for the acceleration a of a cheese travelling at a velocity v in a resisted medium is: $a = g - k.v^p$ where k is some constant depending upon the shape and roughness of the cheese and upon the stickiness (viscosity) of the air, and p is some power depending upon the sort of speed at which the cheese is travelling. Experiments have shown that the terminal velocity of Cheddar cheeses dropping through hot summer's air is about 53 m/s and that the power p is about 2. With these values can you show why k is approximately 0.0035?

Assuming that this acceleration stays roughly constant in each small time interval you can now make up a modified table of velocity v m/s against time t seconds:

t	0	1	2	3	4	5
v	0	9.8	19.26			

How long, approximately, will it take for v to reach 50 m/s? How much does this change if you compute the velocities every 0.1 s instead of every 1 s? (You may need a computer!) What does the graph of v against t look like now? Can you extend the table to include the distance fallen d? What should/does the graph of d against t look like? How far will the Cheddar have fallen when its speed reaches 50 m/s?

A recent newspaper article stated that a Cheddar cheese will reach 99% of its terminal velocity of 53 m/s after about 14 seconds in which time it will have fallen about 570 m. See if you can find a value for p (and the corresponding value of k) for this data.

If the balloon was at a height of 4000 m above the ground when the accidents occurred can you estimate the number of Listeria bacteria in the Camembert at the time the Cheddar hit the ground, and the speed at which the Cheddar was going?

Appendix

The general approach here is sometimes called **dynamic modelling** and is used where quantities change with time – be they physical, biological, economic …

Some simple models, as in Chapters 1, 2, 3, can be solved with ordinary algebra. Models in which increases change continuously usually produce things known as **differential equations** which can sometimes be solved by the techniques of **calculus**.

Models in which increases are assumed to be held constant for short time intervals are called discrete models and produce things known as **difference equations**. Where the continuous change model can be solved by calculus, the discrete model can usually be shown to approximate to the same solution if the time intervals are small. Where the continuous change model cannot be solved by calculus, the numerical approach, such as the one used here, is the only approach left.

The simple approximation we have been using is one often attributed to Leonhard Euler (1707–1783). Better methods have been developed and are usually studied within a branch of mathematics called **numerical analysis.** These are particularly suited for use with a computer.

In fact the model of resisted motion in Chapter 5 can only be solved by calculus for simple values of p such as $p = 1$ or $p = 2$, and the discrete, numerical approach is the only available technique for the data provided.

The model in Chapter 4 is of considerable recent interest since the continuous and discrete approaches differ widely is the rate r (as a decimal) is large – e.g. try $r = 1$, $r = 2$ and $r = 3$; for some values of r the count c becomes **chaotic**. In modelling, the proof of the pudding is in its ability to predict the kind of behaviour that we can observe in nature. Through the chaotic behaviour of simple discrete models like this we can reproduce some sorts of observed behaviour, such as chaotic growth of bacteria or turbulence in fluids, which we had been unable to do previously with continuous models.

7 Mathematics of population and food

Thomas Robert Malthus

1,2,3,4,5,6,7, …
Arithmetic progression

1,2,4,8,16,32,64,128 …
Geometric progression

This is an extract from Thomas Malthus's pamphlet 'An Essay on the Principle of Population As It Affects the Future Improvement of Society'. Thomas Malthus (1766-1834) studied history, poetry, classics and mathematics at Cambridge. He became an Anglican curate in 1797, and the following year anonymously published the first edition of the pamphlet. Several years later he became a professor of political economy, and was elected a Fellow of the Royal Society in 1819.

The essay is a long discussion based on two simple mathematical models: that population growth is a geometric progression and that the growth in the food supply is an arithmetic progression. It was written by a privileged member of a colonialist and male-dominated society, and contains some statements that would not be acceptable today. Also, its somewhat old-fashioned literary style tends to obscure the arguments and conclusions at times.

However, this essay provides an interesting historical example, giving some of the evidence on which Malthus set up his models. It also details his analysis and interpretation of the models and their interaction, and outlines his gloomy prediction that the goal of a good life for all was not attainable without severe restrictions on population growth.

Whether or not the Malthus models are valid, and whether his predictions from them have turned out to be useful, is a matter of judgement and of statistical and scientific investigation. You may have done some of this yourself, in Activity 3.6 of Book 1.

In an inquiry concerning the improvement of society the mode of conducting the subject which naturally presents itself is,
1   to investigate the causes that have hitherto impeded the progress of mankind towards happiness; and
2   to examine the probability of the total or partial removal of these causes in the future.

To enter fully into this question and to enumerate all the causes that have hitherto influenced human improvement would be much beyond the power of an individual. The principal object of the present essay is to examine the effects of one great cause intimately united with the very nature of man; which, though it has been constantly and powerfully operating since the commencement of society, has been little noticed by the writers who have treated this subject. The facts which establish the existence of this cause have, indeed, been repeatedly stated and acknowledged; but its natural and necessary effects have been almost totally overlooked; though probably among these effects may be reckoned a very considerable portion of that vice and misery, and of that unequal distribution of the bounties of nature, which it has been the unceasing object of the enlightened philanthropist in all ages to correct.

The cause to which I allude is the constant tendency in all animated life to increase beyond the nourishment prepared for it.

It is observed by Dr Franklin that there is no bound to the prolific nature of plants or animals but what is made by their crowding and interfering with each other's means of subsistence. Were the face of the Earth, he says, vacant of other plants, it might be gradually sowed and overspread with one kind only, as for instance with fennel; and were it empty of other inhabitants, it might in a few ages be replenished from one nation only, as for instance with Englishmen.

This is incontrovertibly true. Through the animal and vegetable kingdoms nature has scattered the seeds of life abroad with the most profuse and liberal hand, but has been comparatively sparing in the room and the nourishment necessary to rear them. The germs of existence contained in this Earth, if they could freely develop themselves, would fill millions of worlds in the course of a few thousand years. Necessity, that imperious all-pervading law of nature, restrains them within the prescribed bounds. The race of plants and the race of animals shrink under this great restrictive law; and man cannot by any efforts of reason escape from it.

In plants and irrational animals the view of the subject is simple. They are all impelled by a powerful instinct to the increase of their species; and this instinct is interrupted by no doubts about providing for their offspring. Wherever therefore there is liberty, the power of increase is exerted; and the superabundant effects are repressed afterwards by want of room and nourishment.

The effects of this check on man are more complicated. Impelled to the increase of his species by an equally powerful instinct, reason interrupts his career, and asks him whether he may not bring beings into the world for whom he cannot provide the means of support. If he attend to this natural suggestion, the restriction too frequently produces vice. If he hear it not, the human race will be constantly endeavouring to increase beyond the means of subsistence. But as, by that law of our nature which makes food necessary to the life of man, population can never actually increase beyond the lowest nourishment capable of supporting it, a strong check on population, from the difficulty of acquiring food, must be constantly in operation. This difficulty must fall somewhere, and must necessarily be severely felt in some or other of the various forms of misery, or the fear of misery, by a large portion of mankind.

That population has this constant tendency to increase beyond the means of subsistence, and that it is kept to its necessary level by these causes will sufficiently appear from a review of the different states of society in which man has existed. But before we proceed to this review the subject will, perhaps, be seen in a clearer light, if we endeavour to ascertain what would be the natural increase of population if left to exert itself with perfect freedom, and what might be expected to be the rate of increase in the productions of the earth under the most favourable circumstances of human industry.

It will be allowed that no country has hitherto been known where the manners were so pure and simple, and the means of subsistence so abundant, that no check whatever has existed to early marriages from the difficulty of providing for a family, and that no waste of the human species has been occasioned by vicious customs, by towns, by unhealthy occupations, or too severe labour. Consequently, in no state that we have yet known has the power of population been left to exert itself with perfect freedom.

Whether the law of marriage be instituted or not, the dictates of nature and virtue seem to be an early attachment to one woman; and where there were no impediments of any kind in the way of an union to which such an attachment would lead, and no causes of depopulation afterwards, the increase of the human species would be evidently much greater than any increase which has hitherto been known.

In the Northern States of America, where the means of subsistence have been more ample, the manners of the people more pure, and the checks to early marriages fewer than in any of the modern states of Europe, the population has been found to double itself, for above a century and a half successively, in less than twenty-five years. Yet, even during these periods, in some of the towns the deaths exceeded the births, a circumstance which clearly proves that, in those parts of the country which supplied this deficiency, the increase must have been much more rapid than the general average.

In the back settlements, where the sole employment is agriculture, and vicious customs and unwholesome occupations are little known, the population has been found to double itself in fifteen years. Even this extraordinary rate of increase is probably short of the utmost power of population. Very severe labour is requisite to clear a fresh country; such situations are not in general considered as particularly healthy; and the inhabitants, probably, are occasionally subject to the incursions of the Indians, which may destroy some lives, or at any rate diminish the fruits of industry.

According to a table of Euler, calculated on a mortality of one to thirty-six, if the births be to the deaths in proportion of three to one, the period of doubling will be only twelve years and four fifths. And this proportion is not only a possible supposition, but has actually occurred for short periods in more countries than one.

Sir William Petty supposes a doubling possible in so short a time as ten years.

But, to be perfectly sure that we are far within the truth, we will take the slowest of these rates of increase, a rate in which all concurring testimonies agree, and which has been repeatedly ascertained to be from procreation only.

It may safely be pronounced, therefore, that population, when unchecked, goes on doubling itself every twenty-five years, or increases in a geometrical ratio.

The rate according to which the productions of the earth may be supposed to increase it will not be so easy to determine. Of this, however, we may be perfectly certain – that the ratio of their increase in a limited territory must be of a totally different nature from the ratio of the increase of population. A thousand millions are just as easily doubled every twenty-five years by the power of population as a thousand. But the food to support the increase from the greater number will by no means be obtained with the same facility. Man is necessarily confined in room. When acre has been added to acre till all the fertile land is occupied, the yearly increase of food must depend upon the melioration of the land already in possession. This is a fund which, from the nature of all soils instead of increasing, must be gradually diminishing. But population, could it be supplied with food, would go on with unexhausted vigour; and the increase of one period would furnish the power of a greater increase the next, and this without any limit.

From the accounts we have of China and Japan, it may be fairly doubted whether the best-directed efforts of human industry could double the produce of these countries even once in any number of years. There are many parts of the globe, indeed, hitherto uncultivated and almost unoccupied, but the right of exterminating, or driving into a corner where they must starve, even the inhabitants of these thinly-peopled regions, will be questioned in a moral view. The process of improving their minds and directing their industry would necessarily be slow; and during this time, as population would regularly keep pace with the increasing produce, it would rarely happen that a great degree of knowledge and industry would have to operate at once upon rich unappropriated soil. Even where this might take place, as it does sometimes in new colonies, a geometrical ratio increases with such extraordinary rapidity that the advantage could not last long. If the United States of America continue increasing, which they certainly will do, though not with the same rapidity as formerly, the Indians will be driven further and further back into the country, till the whole race is ultimately exterminated and the territory is incapable of further extension.

These observations are, in a degree, applicable to all the parts of the Earth where the soil is imperfectly cultivated. To exterminate the inhabitants of the greatest part of Asia and Africa is a thought that could not be admitted for a moment. To civilize and direct the industry of the various tribes of Tartars and Negroes would certainly be a work of considerable time, and of variable and uncertain success.

Europe is by no means so fully peopled as it might be. In Europe there is the fairest chance that human industry may receive its best direction. The science of agriculture has been much studied in England and Scotland, and there is still a great portion of uncultivated land in these countries. Let us consider at what rate the produce of this island might be supposed to increase under circumstances the most favorable to improvement.

If it be allowed that by the best possible policy, and great encouragement to agriculture, the average produce of the island could be doubled in the first twenty-five years, it will be allowing, probably, a greater increase than could with reason be expected.

In the next twenty-five years it is impossible to suppose that the produce could be quadrupled. It would be contrary to all our knowledge of the properties of land. The improvement of the barren parts would be a work of time and labour; and it must be evident, to those who have the slightest acquaintance with agricultural subjects, that in proportion as cultivation is extended, the additions that could yearly be made to the former average produce must be gradually and regularly diminishing. That we may be the better able to compare the increase of population and food, let us make a supposition which, without pretending to accuracy, is clearly more favourable to the power of production in the earth than any experience we have had of its qualities will warrant.

Let us suppose that the yearly additions which might be made to the former average produce, instead of decreasing, which they certainly would do, were to remain the same; and that the produce of this island might be increased every twenty-five years by a quantity equal to what it at present produces. The most enthusiastic speculator cannot suppose a greater increase than this. In a few centuries it would make every acre of land in the island like a garden.

If this supposition be applied to the whole Earth, and if it be allowed that the subsistence for man which the earth affords might be increased every twenty-five years by a quantity equal to what it all present produces, this will be supposing a rate of increase much greater than we can imagine that any possible exertions of mankind could make it.

It may be fairly pronounced, therefore, that considering the present average state of the Earth, the means of subsistence, under circumstances the most favourable to human industry, could not possibly be made to increase faster than in an arithmetical ratio.

The necessary effects of these two different rates of increase, when brought together, will be very striking. Let us call the population of this island eleven millions; and suppose the present produce equal to the easy support of such a number. In the first twenty-five years the population would be twenty-two millions, and the food being also doubled, the means of subsistence would be equal to this increase. In the next twenty-five years the population would be forty-four millions, and the means of subsistence only equal to the support of thirty-three millions. In the next period the population would be eighty-eight millions and the means of subsistence just equal to the support of half that number. And at the conclusion of the first century the population would be a hundred and seventy-six millions, and the means of subsistence only equal to the support of fifty-five millions, leaving a population of a hundred and twenty-one millions totally unprovided for.

Taking the whole Earth instead of this island, emigration would of course be excluded; and, supposing the present population equal to a thousand millions, the human species would increase as the numbers, 1, 2, 4, 8, 16, 32, 64, 128, 256, and subsistence as 1, 2, 3, 4, 5, 6, 7, 8, 9. In two centuries the population would be to the means of subsistence as 256 to 9; in three centuries, as 4096 to 13; and in two thousand years the difference would be almost incalculable.

In this supposition no limits whatever are placed to the produce of the Earth. It may increase forever, and be greater than any assignable quantity; yet still, the power of population being in every period so much superior, the increase of the human species can only be kept down to the level of the means of subsistence by the constant operation of the strong law of necessity, acting as a check upon the greater power.

Of the general checks to population, and the mode of their operation

The ultimate check to population appears then to be a want of food arising necessarily from the different ratios according to which population and food increase. But this ultimate check is never the immediate check, except in cases of actual famine.

The immediate check may be stated to consist in all those customs, and all those diseases, which seem to be generated by a scarcity of the means of subsistence; and all those causes, independent of this scarcity, whether of a moral or physical nature, which tend prematurely to weaken and destroy the human frame.

These checks to population, which are constantly operating with more or less force in every society, and keep down the number to the level of the means of subsistence, may be classed under two general heads – the preventive, and the positive checks.

The preventive check, as far as it is voluntary, is peculiar to man, and arises from that distinctive superiority in his reasoning faculties which enables him to calculate distant consequences. The checks to the indefinite increase of plants and irrational animals are all either positive, or, if preventive, involuntary. But man cannot look around him and see the distress which frequently presses upon those who have large families; he cannot contemplate his present possessions or earnings, which he now nearly consumes himself, and calculate the amount of each share, when with very little addition they must be divided, perhaps among seven or eight, without feeling a doubt whether, if he follow the bent of his inclinations, he may be able to support the offspring which he will probably bring into the world. In a state of equality, if such can exist, this would be the simple question. In the present state of society other considerations occur. Will he not lower his rank in life, and be obliged to give up in great measure his former habits? Does any mode of employment present itself by which he may reasonably hope to maintain a family? Will he not at any rate subject himself to greater difficulties and more severe labour than in his single state? Will he not be unable to transmit to his children the same advantages of education and improvement that he had himself possessed? Does he even feel secure that, should he have a large family, his utmost exertions can save them from rags and squalid poverty, and their consequent degradation in the community? And may he not be reduced to the grating necessity of forfeiting his independence, and of being obliged to the sparing hand of charity for support?

These considerations are calculated to prevent, and certainly do prevent, a great number of persons in all civilized nations from pursuing the dictate of nature in an early attachment to one woman.

If this restraint does not produce vice, it is undoubtedly the least evil that can arise from the principle of population. Considered as a restraint on a strong natural inclination, it must be allowed to produce a certain degree of temporary unhappiness, but evidently slight compared with the evils which result from any of the other checks to population, and merely of the same nature as many other sacrifices of temporary to permanent gratification, which it is the business of a moral agent continually to make.

When this restraint produces vice, the evils which follow are but too conspicuous. A promiscuous intercourse to such a degree as to prevent the birth of children seems to lower, in the most marked manner, the dignity of human nature. It cannot be without its effect on men, and nothing can be more obvious than its tendency to degrade the female character and to destroy all its most amiable and distinguishing characteristics. Add to which, that among those unfortunate females with which all great towns abound more real distress and aggravated misery are, perhaps, to be found, than in any other department of human life.

When a general corruption of morals with regard to the sex pervades all the classes of society, its effects must necessarily be to poison the springs of domestic happiness, to weaken conjugal and parental affection, and to lessen the united exertions and ardour of parents in the care and education of their children – effects which cannot take place without a decided diminution of the general happiness and virtue of the society; particularly as the necessity of art in the accomplishment and conduct of intrigues and in the concealment of their consequences necessarily leads to many other vices.

The positive checks to population are extremely various, and include every cause, whether arising from vice or misery, which in any degree contributes to shorten the natural duration of human life. Under this head, therefore, may be enumerated all unwholesome occupations, severe labour and exposure to the seasons, extreme poverty, bad nursing of children, great towns, excesses of all kinds, the whole train of common diseases and epidemics, wars, plague, and famine.

On examining these obstacles to the increase of population which I have classed under the heads of preventive and positive checks, it will appear that they are all resolvable into moral restraint, vice and misery.

Of the preventive checks, the restraint from marriage which is not followed by irregular gratification may properly be termed moral restraint.[1]

Promiscuous intercourse, unnatural passions, violations of the marriage bed, and improper arts to conceal the consequences of irregular connections are preventive checks that clearly come under the head of vice.

Of the positive checks, those which appear to arise unavoidably from the laws of nature may be called exclusively misery, and those which we obviously bring upon ourselves, such as wars, excesses, and many others which it would be in our power to avoid, are of a mixed nature. They are brought upon us by vice, and their consequences are misery.

The sum of all these preventive and positive checks taken together forms the immediate check to population; and it is evident that in every country where the whole of the procreative power cannot be called into action, the preventive and the positive checks must vary inversely as each other; that is, in countries either naturally unhealthy or subject to a great morality, from whatever cause it may arise, the preventive check will prevail very little. In those countries, on the contrary, which are naturally healthy, and where the preventive check is found to prevail with considerable force, the positive check will prevail very little, or the mortality be very small.

In every country some of these checks are with more or less force in constant operation; yet notwithstanding their general prevalence, there are few states in which there is not a constant effort in the population to increase beyond the means of subsistence. This constant effort as constantly tends to subject the lower classes of society to distress, and to prevent any great permanent melioration of their condition.

---

[1]It will be observed that I here use the term *moral* in its most confined sense. By *moral* restraint I would be understood to mean a restraint from marriage from prudential motives, with a conduct strictly moral during the period of this restraint; and I have never intentionally deviated from this sense. When I have wished to consider the restraint from marriage unconnected with its consequences, I have either called it prudential restraint, or a part of the preventative check, of which indeed it forms the principal branch. In my review of the different stages of society I have been accused of not allowing sufficient weight in the prevention of population to moral restraint; but when the confined sense of the term, which I have here explained, is adverted to, I am fearful that I shall not be found to have erred much in this respect. I should be very glad to believe myself mistaken.

8 Evaluate risk in plant design

Trevor Kletz

Hazard analysis is growing in importance. Expensive fires and explosions – and the prospect of worse ones – means more attention must focus on pre-planning disasters, and on taking remedial action beforehand.

A major application of probability is to the analysis of risk in business, industry and personal life. This article discusses some aspects of the use of risk analysis from the point of view of a designer of chemical plants. The ideas can also be applied to a wide variety of cicumstances including the risk involved with a particular portfolio of shares, and the way in which a doctor weighs up the circumstances for an individual patient.

In any industry, particularly a complex one like the hydrocarbon processing industry, it is impossible to remove every risk. So how do we decide which risks should be dealt with first and which risk can be left, at least for the time being? In short, how do we allocate our resources?

I have tried to show this in Figure 8.1. The horizontal axis shows expenditure on safety over and above that necessary for a workable plant. The vertical axis shows the money we get back in return. In the left-hand area safety is good business – by spending money on safety, apart from preventing injuries our plants run for longer and we make more profit.

In the next area safety is poor business – we get some money back from our safety expenditure but not as much as we would get by spending our money in other ways.

We then move into an area where safety is bad business but good humanity, and finally into an area where we are spending so much on safety that we are going out of business. Our products are becoming so expensive that no one will buy them. What we have to do is to decide where to draw the line between the last two areas.

Usually this is a qualitative judgement but it is often possible to make it quantitative.

Hazard analysis is the name given to these attempts to apply quantitative methods to safety problems. Before discussing these methods it may be useful to distinguish between hazard analysis and operability studies, as they are often confused.

Operability studies, also called hazard and operability studies or "Hazops," are a means of *identifying* hazards or other problems by using a series of guide words such as NONE, MORE OF, LESS OF and so on.[1,2] Other means, such as check lists, can be used to identify problems but their disadvantage is that items not in the list do not get checked. Operability studies are more "open-ended." Of course, some problems are obvious; if we mix ethylene and air an explosion may occur if they are mixed in the wrong proportions. We do not need an operability study or checklist to tell us we have a potential hazard.

How serious? When we have identified our hazards or other problems, we have to decide whether they are so serious that we should do something about them, or whether we can leave them, temporarily. Sometimes the decision is obvious. Sometimes it can be made by referring to a Code of Practice or previous experience. Sometimes it is made by judgement. Sometimes, to aid our judgement, we attempt to calculate the probability that the hazard will occur and compare with a target or specification figure. This is called hazard analysis. It can vary from a quick mental sum to a detailed study taking many weeks.

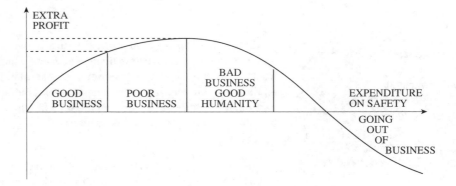

Figure 8.1

This illustration speaks to the question, "Which risks do we try to eliminate first, for optimum over-all protection of life and property?"

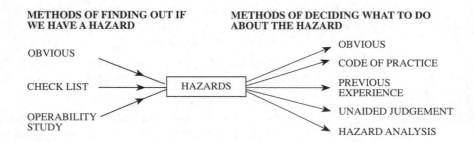

Figure 8.2

Hazard analysis, whether done quickly on a "shoot-from -the-hip basis", or in a detailed study of several moths, will involve these considerations

Figure 8.2 may make the difference clear.

In the development of a project the operability study comes first. We identify our hazards and then decide what to do abut them. However, if there is an obvious major hazard we may start on the hazard analysis before the operability study is carried out.

The term "risk analysis" should be avoided. It is used to describe a method of estimating the *commercial* risks of a project.[3]

Hazard-analysis stages. Any hazard analysis, however simple, consists of several steps:

1   Estimating the probability the hazard will occur (hazard)

2   Estimating the consequence the hazard

3   Comparing the results of **1** and **2** with a target or specification to decide whether or not action to reduce the probability of occurrence or minimise the consequences is desirable or whether the risk should be left, at least for the time being.

Separate targets or specifications are needed for:
* risks to profits in circumstances in which injury is unlikely
* risks to employees
* risks to members of the public. A higher standard is required than for employees as a member of the public may have risks imposed on him against his will.

Sometimes past experience will tell us the hazard. We may know that a particular pump leaks four times per year or a tank is overfilled five times per year. Often we have no experience and have to estimate a hazard rate. The section on *Examples of how hazard rates can be estimated* at the end of this article shows by example how this can be done.

More detailed examples are given in literature citations 1, 4, 5 and 6.

Estimation of consequences

Depending on the nature of the hazard this may be expressed as the probability that someone will be killed or injured or it may be an estimate of the expected damage or loss of profit. When the probability that someone will be killed or injured is very small the target hazard rate will be higher than when the probability of death or injury is significant.

The most difficult part of hazard analysis is deciding on the *target* or *specification* by which to judge hazards – when is the probability of an accident so low that we should ignore it? There are many papers on methods of estimating the probability that an accident will occur but only a handful [4-9] on methods of deciding what hazard rate is acceptable. Yet it is no use estimating hazard rates unless we can use them as an aid to decision making.

So here I will discuss various criteria proposed for deciding whether or not hazards should be dealt with as a matter of priority or whether they can be deferred.

Risks to profits

This is dealt with first, not because profits are more important than injuries, but because the problem is simpler and is best got out of the way.

If the risk to life is so high that the risk must be reduced as a matter of priority, then the economic loss is of no concern – the risk must be dealt with anyhow as a matter of priority. If, however, the risk to life is so low that it can be tolerated in the short term, the economic loss should be calculated to see if this provides a reason for removing the risk.

The procedure is as follows:

1   The probability that the dangerous occurrence will occur is estimated. Let us suppose it is once in 1,000 years.

2   The loss of profit that will result is estimated. Consequential losses should be included as well as the cost of repairing the damage. Suppose the loss is $2 million.

3   From **1** and **2**, the average annual loss, L, is calculated. In this case it is $2000.

4   The net cost, C (after allowing for tax allowances) of reducing the probability that the dangerous occurrence will occur is calculated. Let us suppose it will cost $15,000 to prevent it happening.

5   The cost, C, is converted into an annual value (maintenance, depreciation and interest on capital). In the example considered it will be about $5,000.

Net worth. If this annual value is less than L the expenditure of C is justified. In the example, the annual cost is $5,000 and the average annual cost L is $2,000 so the expenditure is not justified.

Strictly speaking, since most of the "premium" has to be paid now and the loss, on average, only after some years, the loss should be converted to its "net present worth." The data are not, however, usually accurate enough to make this refinement worthwhile.

In calculating the loss, the costs of any injuries or fatalities (compensation, legal costs, costs of investigation, etc.) can be ignored. They are usually negligible in comparison with the cost of the damage and loss of production.

Risks to employees

A method that has been widely used is to compare risks to life per hour of exposure and then concentrate resources on those which exceed a specified value.

The risk to life is expressed in deaths per 10^8 exposed hours, i.e. hours at work. We are familiar with the lost-time accident frequency rate which is expressed at lost-time accidents per 10^6 exposed hours, 10^6 hours being used as it is a working life time. In discussing fatal accidents, in order to avoid using small fractions, a fatal accident frequency rate (FAFR) of fatalities per 10^8 hours is used; it is the number of deaths from industrial injury expected in a group of 1,000 men during their working lives.

Fatal accident frequency rates for a number of industries and occupations have been derived by Sowby[10] and Pochin[11] and some are quoted in Table 8.1.

In considering the last three figures in Table 8.1 it must be remembered that air crews are exposed for less than the normal 40 hours per week and boxers and jockeys for very much less.

Individual firms or factories do, of course, have accident rates far below the average for their industries. It has been said that if all companies could achieve the safety record of the best, accidents would be cut by 80%.

The figures for the chemical industry pre-dates Flixborough. It will be about 5 if Flixborough is averaged over 10 years.

Clothing and footwear	0.15
Vehicles	1.3
Chemical industry	4
British industry (i.e. all premises covered by the Factory Act)	4
Metal manufacture and shipbuilding	8
Agriculture	10
Fishing	36
Coal mining	14
Railway shunters	45
Construction workers	67
Air crew	250
Professional boxers	7,000
Jockeys (flat racing)	50,000

Table 8.1

FAFR for various industries

For comparison, FAFRs for some other activities are given in Table 8.2, all expressed as deaths per 10^8 exposed hours.

Staying at home	3
Travelling by bus	3
Travelling by train	5
Travelling by car	57
Pedal cycling	96
Travelling by air	240
Moped riding	260
Motor scooter driving	310
Motor cycling	660
Canoeing	1,000
Rock climbing	4,000

Table 8.2

FAFR for non-industrial activities

The figure for staying at home assumes that half the population are at home for 8 hours per day and half for 16 hours per day, i.e. time spent away from home or in bed is excluded. Most of the accidents happen to children and old people; the risk to an able-bodied man is lower – about one.

The frequency rates of fatalities in Table 8.3 per 10^8 hours for other causes of death have been calculated by Sowby assuming 24 hours exposure per day, except for accidents for which 16 hours exposure per day is assumed.

Chemical industry

It is seen that the FAFR for the chemical industry is about the same as that for British industry as a whole despite the fact that the chemical industry handles flammable, explosive, toxic and corrosive materials. Working in the chemical industry is far safer than mining or construction. Nevertheless our aim is, and must be, to reduce our accident rate further. Our ultimate aim must be to make it zero. But we cannot do everything at once. We should therefore concentrate our resources first on the greatest risks.

All causes (male and female)	133
Cardiovascular diseases (male and female)	61
All malignant neoplasms (male)	23
Respiratory disease (male)	22
Lung cancer (male)	10
Stomach cancer (male)	4
All accidents (male)	9

Table 8.3

FAFR for other causes of death

Furthermore, we are all at risk all the time, whatever we do, even staying at home. We accept the risks when we consider that by doing so something worthwhile is achieved. We go rock climbing or sailing or we smoke because we consider that the pleasure is worth the risk. We tke jobs as airline pilots or soldiers or we become missionaries among cannibals because we consider that the pay, or the interest of the job, or the benefit it brings to others, makes the risk worthwhile.

At work there is likely to be a slight risk, whatever we do to remove known risks. By accepting this risk we earn our living and we make goods that enable ourselves and others to lead a fuller life.

Non-industry related risks

The average total risk to life in the chemical industry is four fatalities per 10^8 exposed hours, an FAFR of four. About half of this is made up of risks unconnected with the nature of the industry, such as falling downstairs or getting run over by a vehicle. About half is made up of special chemical risks, such as fire, explosion and toxicity. If we are sure that we have identified all the special risks attached to a particular job, we set as our specification that the man doing the job should not be exposed to an FAFR, for these special risks, greater than two. We will eliminate or reduce, as a matter of priority, any such risks on new or existing plants.

On most plants we are not sure that we have identified all the special risks. So we set as our specification that any single one, considered in isolation, should not expose an employee to an FAFR greater than 0.4. We will eliminate or reduce, as a matter of priority, any hazard, on a new or existing plant, that exceeds this figure. We are thus assuming that there are about five special risks on a typical plant.

Voluntary			Involuntary		
Activity	Risk of Death per person per year	Ref	Activity	Risk of Death per person per year	Ref
Smoking (20 cigarettes per day)	500×10^{-5}	11	Run over by road vehicle (USA)	500×10^{-7}	16
Drinking (one bottle wine per day)	7.5×10^{-5}	11	Run over by road vehicle (UK)	600×10^{-7}	..
Football	4×10^{-5}	11	Floods (USA)	22×10^{-7}	14
Car racing	120×10^{-5}	11	Earthquake (California)	17×10^{-7}	14
Rock climbing	4×10^{-5}	11	Tornadoes (Mid-West, USA)	22×10^{-7}	14
Car driving	17×10^{-5}	11	Storms (USA)	8×10^{-7}	14
Motor racing	$2,000 \times 10^{-5}$	11	Lightning (UK)	1×10^{-7}	6
Taking contraceptive pills	2×10^{-5}	8	Falling aircraft (USA)	1×10^{-7}	..
			Falling aircraft (UK)	0.2×10^{-7}	8
			Explosion of pressure vessel (USA)	0.5×10^{-7}	16
			Release from atomic power station		
			At site building (USA)	1×10^{-7}	16
			At 1 km (UK) 1×10^{-7}	..	
			Flooding of dikes (Holland)	1×10^{-7}	17
			Bites of venomous creatures (UK)	2×10^{-7}	8
			Transport of petrol/chemicals (USA)	0.5×10^{-7}	16
			Transport of petrol/chemicals (UK)	0.2×10^{-7}	..
			Leukemia	800×10^{-7}	8
			Influenza	$2,000 \times 10^{-7}$	8
			Meteorite	6×10^{-11}	16
			Supernovae explosion cosmic rays	$10^{-8} - 10^{-11}$	16

Table 8.4

Voluntary and involuntary risks compared

Experience has shown that the costs of doing this, though often substantial, are not unbearable. They involve us in expenditure which some of our competitors do not incur. Some of the extra expenditure can be recouped in lower insurance premiums; some can be recouped by the greater plant reliability which safety measures often produce; the rest is a self-imposed 'tax' which has to be balanced by greater efficiency.

FAFR rates

If a job is manned by one man for 2,000 hours per year an FAFR of 0.4 is equivalent to one fatality every 125,000 years. If the job is manned continuously on shifts, then it is equivalent to one fatality every 30,000 years.

As mentioned above, an expression often used is the hazard rate, the rate at which dangerous conditions arise on a plant. If every time a dangerous condition arises the person doing the job is killed, then the specification hazard rate for the plant (corresponding to an FAFR of 4.0) is 1 in 30,000 running years or 3×10^{-5} per running year. If, on the other hand, it is estimated that there is a chance of 1 in 10 that he will be killed when the dangerous condition arises, the specification hazard rate is 1 in 3,000 running years or 3×10^{-4} per running year.

Analysis guideline

There are a few guidelines which should be borne in mind in applying Hazard Analysis.

1   We should assume that the job is done by the same man every day, or by the same four men if it is a shift job. We cannot spread the risk by employing a different man every day to do a hazardous job.
2   On the other hand, the risks to which a man is exposed from a particular hazard may be averaged over the working day, to give an FAFR of 0.4 or below. Obviously some parts of a job are more risky than others. It is impracticable to specify that each moment's task if undertaken continuously, must give an FAFR below 0.4. We are willing, for example to cross a busy road, but would hardly be willing to spend our days doing so.
3   In some hazards the risk has to be spread over many employees. The risk to each of them is small but the total risk may be significant. In these cases we assume that the man killed is always the same man. (We may call this man Jonah, as, by jumping into the sea, Jonah took the risk of drowning away from the sailors and exposed himself to a much greater risk). For example, in considering the possibility that the roof of a tank will fly off and kill a man on the ground, any one of a large number of men might be killed. We therefore assume that if the roof lands on anyone, it will always be Jonah and that, furthermore, Jonah is the man whose duty it is to go on to the tank roof.

Hazard analysis has been pioneered in the chemical industry. It can, of course, be applied to other industries. But the specification FAFR will be different, lower for a safe industry, initially higher for a hazardous industry such as coal mining.

Use of the specification

On a particular plant it was found that it was possible for a refrigerated gas to come into contact with a mild steel pipe line. If this occurred the pipe line might have

cracked. Two methods of overcoming the hazard were considered.
a   Replacing the mild steel by stainless steel at a considerable cost, or
b   Improving the alarm and trip system at one-quarter of the cost.

As redesigned the system contains three independent layers of protection:
1   There is a high-level alarm on a catchpot.
2   If this fails to operate or is ignored, an independent high-level trip closes a valve in the inlet line to the catchpot.
3   If this fails to operate, a low-temperature trip on the overhead line from the catchpot closes a valve in the overhead line. (The overhead line is made of stainless steel but leads to a mild steel line.)

Calculations

Using data on the reliability of the various components of the trip system and assuming that the operator will ignore the initial alarm signal on one occasion out of four the "fractional dead time" of the whole system (that is, the fraction of the time it is not operating) was calculated. This depends on the frequency with which the trips are tested.

The "demand rate" on the system, the number of times per year it is called on to operate, was estimated from previous experience.

The failure rate of the whole system was then estimated as once in 10,000 years or once in 2,500 years for the whole plant which contained four similar systems.

It was assumed that one-tenth of the occasions on which the cold gas reached mild steel would result in a leak, an explosion and a fatality – almost certainly an overestimate. A fatality will then occur once in 25,000 years giving an FAFR of 0.4. It was therefore agreed that the control system was satisfactory and it was not necessary to replace the mild steel by stainless steel.

It might be argued that the latter course is preferable as it is 100 percent safe: the FAFR is zero. Had we done this we would have spent on reducing a very slight hazard money and other resources which would be better spent on dealing with greater hazards.

Simpler criteria

It is not always necessary to estimate the FAFR. Simpler criteria can often be used.

For example, suppose we intend to use an instrumented protective system instead of a relief valve, perhaps in order to reduce the size of the flare system required. A high-pressure switch will detect a rise in pressure in a vessel and will close a valve which isolates the source of the pressure. If the vessel is a distillation column, and the rise in pressure is due to the loss of reflux, then the pressure switch would close a valve in the steam line to the reboiler.

How do we decide on the reliability required in the protective system? One possible method would be to estimate the consequences of the protective system failing to work. The vessel would be overpressured. It might fail; we could estimate the probability. Some of the contents could leak out; we could estimate how much. If flammable they might ignite; we could estimate the probability. Someone might be in the area; we could estimate the probability. He might be killed; we could estimate the probability. Finally we could estimate the FAFR.

An easier way

A simpler method of calculation is, however, possible. Relief valves are widely used and their reliability, although not 100 per cent, is generally regarded as acceptable.

There is no demand for more reliable devices. The protective system used in place of a relief valve should therefore be just as reliable, preferably rather more reliable as we are using something new in place of something well established. Also there is a difference in the mode of failure of relief valves and protective systems; a relief valve which fails to lift at, say, twice the set pressure may still lift at a higher pressure. The same is not true of a protective system. We therefore propose that a protective system used in place of a relief valve should have a failure rate ten times lower.

> For more details see literature citations 12 and 13.

Risks to population

When considering hazards which might affect the public at large, the level of risk which can be considered tolerable, even in the short term, is much lower. A man chooses to work for a particular employer or in a particular industry. Unless he chooses a high-hazard occupation, the risk to which he is exposed is not much greater than if he had stayed at home. On the other hand, a member of the public may have risks imposed on him without his consent.

> Other examples of the application of hazard analysis to risks to employees are given in literature citations 1, 4, 5, 6 and 7. Literature citation 11 discusses risks due to the long-term effects of toxic chemicals.

Starr [14,15] has pointed out that we accept voluntarily a number of risks such as driving, flying, and smoking, which expose us to a risk of death of 10^{-5} or more, sometimes a lot more, per person per year (FAFR 0.1). We also accept, with little or no complaint, a number of involuntary risks which expose us to a risk of death of about 10^{-7} or less per person per year (FAFR 0.001).

The record

Table 8.4 lists some of these voluntary and involuntary risks. The figures are, of course, only approximate and may have been calculated using different assumptions. Furthermore, some of the comparatively small number of workers in this field copy from each other, so any error, once introduced, is repeated in various papers and acquires an aura of authenticity. Any individual figure should therefore be treated with reserve, and if it is used in a calculation, checked in the original sources. Nevertheless, the figures do demonstrate Starr's point.

We accept very high risks voluntarily; we accept other risks, imposed on us without our leave, if they are sufficiently small. It would be possible to do something about the involuntary risks listed in Table 8.4 if there was sufficient pressure from the press and public, but on the whole there is no such pressure. The risk of being struck by lightning or falling aircraft is so small that we accept the occasional death without complaint.

Cars, disasters

From road vehicles we accept very high risks, presumably the advantages to us from road vehicles are clear and obvious. From natural disasters we accept risks of about 10^{-6} per person per year; from man-made events, except road transport, we seem to accept about 10^{-7} per person per year.

Leukemia and influenza have been included in the list of involuntary risks as examples of risks we do not readily accept; there is pressure for something to be done. Most people would support action to reduce their incidence, but would consider the other involuntary risks hardly worth bothering about. There is some concern about the road transport of chemicals, but this is based, at least in part, on the fear that more serious incidents might occur.

Derivation of a risk specification

We thus have a basis for assessing risks to the public at large from industrial activity. If the average risk to those exposed is less than 10^{-7} per person per year, the risk should be accepted, at least in the short-term, and resources should not be allocated to its reduction.

For many of the risks in Table 8.4 the whole population of a country is at risk but industrial risks on the whole affect only people living within a few kilometres of the factory. Industrial risks cannot therefore be averaged over the whole population, but only over those who are in some degree at risk.

Those living near an airport are at greater risk than those living further away; nevertheless, the average risk is accepted. Similarly, those living near a factory are at greater risk than those living further out, but an average risk should be considered acceptable if it does not exceed 10^{-7} per person per year. Some limitation on the maximum risk of death to which any individual is exposed is desirable and perhaps this should be fixed at 10^{-5} to per person per year.

A risk of 10^{-7} per person per year is, of course, extremely low. It may be explained to non-technical people, who may be unused to probabilities, as follows: suppose all sources of death were removed excepting that resulting from a particular industrial activity. Then all the people living near the factories concerned would have an average lifetime of 10,000,000 years.

Example, use of specification

Large inventories of liquefied toxic gases such as ammonia are best stored refrigerated, at low pressures.[18] However, many existing installations remain in operation. Many of these have been subjected to a detailed hazard analysis in which all circumstances which could lead to a massive release are identified and the size and probability of such a release estimated. The probability of gas reaching residential areas is then calculated, taking wind direction and strength into account. Finally, an attempt is made to calculate the chance that members of the public might be killed.

Many of these studies have shown that the risk to life for members of the public is less than 10^{-7} per person per year, and that therefore no action need be taken to reduce the risk. This is in line with the pragmatic view that there is no record in the United Kingdom (and very few elsewhere in the world) of members of the public being killed by a release of toxic gas from a storage area.

High public risk

One study of the type described did show that the risk to the public was high. The storage vessel was located alongside a works road on which cranes regularly moved. The chance of a crane collapsing on the storage vessel was estimated and the consequent risk to the public was found to be too high. A plan was prepared to protect the vessel with metal girders, at considerable cost. But it was found possible to manage without the vessel and it was taken out of service.

A somewhat similar study has been described by Dicken.[9] All the circumstances which could lead to an emission of chlorine from a plant are identified and the size and probability of a release estimated. The concentration of chlorine at the plant boundary is then estimated and compared with target figures. A "nuisance" is considered acceptable once a year, a release causing "some distress" is considered acceptable once in ten years, and a release which could lead to "personal injury or risk to life" is considered acceptable once in 100 years. All these expressions are quantified in terms of concentration and duration. The last category is roughly equivalent to a risk of 10^7 per person per year for the population living near the plant.

For other examples of the use of the specification see literature citation 7.

Use of simpler criteria

As with risks to employees, simpler methods of calculation can often be used.

An intermediate product was carried 200 miles by road for further processing. The intermediate was in the form of an aqueous solution and was harmless. But money was being spent to transport water. It was therefore proposed to transport instead an alternative intermediate which was water-free but corrosive. The total quantity to be transported would be reduced by over 80 percent. The problem that had to be answered was whether the risk to the public from the transport of a hazardous chemical was so low that it should be accepted, bearing in mind that a safer, though bulkier material could be transported instead. It was assumed that the transportation would be carried out in vehicles of the highest quality by well-trained drivers.

Using average figures for the number of people killed in ordinary road accidents and in accidents involving chemicals it was possible to show that reducing the volume of material to be transported by five-sixths would, on average, save one life every 12 years, even allowing for the fact that an accident involving a tanker of corrosive chemicals is slightly more likely to result in a fatality than an accident involving a tanker of harmless material.

Examples of how hazard rates can be estimated

Relief Valves

According to the UK Atomic Energy Authority relief valves develop faults which prevent them opening when required at a rate of 0.005 faults per year. Our own experience suggests a rather higher rate, say, 0.01 faults per year, i.e. each valve will develop a fault which will prevent it opening when required once in 100 years.

Assume each relief valve is tested once per year. On average, failure will occur half-way between tests and the relief valve will be ineffective until the next test, i.e. for six months in 100 years or for 0.5% of the time. We say the fractional dead time or FDT is 0.5% or 0.005.

Suppose the "demand rate" on the relief valve, i.e. the frequency with which it is called to operate, is once in five years or 0.2 per year.

The "hazard rate", the rate at which the vessel is over pressured, is the rate at which demands coincide with the dead time. Hence:

$$\text{Hazard rate} = \text{Demand rate} \times \text{FDT}$$
$$= \quad 0.2 \text{ per year} \times 0.005$$
$$= \quad 0.001 \text{ per year or once in } 1,000 \text{ years.}$$

The vessel will not, of course, burst on every occasion that it is over-pressured. It may withstand the excess pressure or it may leak at a joint. We have defined as a hazard any failure which causes a vessel to be subjected to a pressure more than 10% above its relief valve setting.

If relief valves are tested every two years, the FDT becomes 1% and the hazard rate becomes once in 500 years.

Simple Trips

Depending on the detailed design, a simple trip consisting of a pressure switch linked to a single motor valve will develop faults which prevent it operating on demand once every year or two, say, once in 18 months.

If the trip is *thoroughly* tested every week, on average it will be ineffective for $3\frac{1}{2}$ days over $1\frac{1}{2}$ years, or for 0.64% of the time. The FDT is 0.0064.

Assuming as before a demand rate of once in five years or 0.2 per year, then:

$$\text{Hazard rate} \quad = \quad \text{Demand rate} \times \text{FDT}$$
$$= \quad 0.2 \text{ per year} \times 0.0064$$
$$= \quad 0.001\ 28 \text{ per year or once in } 750 \text{ years.}$$

If the trip is tested monthly, the FDT becomes 0.02 and the hazard rate becomes once in 180 years.

If the trip is tested annually, the FDT becomes 0.3 and the hazard rate becomes once in 15 years, not much better than the hazard rate for no trip at all (once in five years).

This calculation is not quite accurate. The correct formula is

$$\text{Hazard rate} = f\left[1 - \exp\left(-\frac{DT}{2}\right)\right]$$

where f = failure rate, T = proof test interval, D = demand rate.

When $DT/2$ is small this reduces to: $\quad \text{Hazard rate} = \frac{1}{2}fDT$

Duplicated Trips

Suppose we have two trips and test each one weekly. The fractional dead time of the combined system, assuming they are tested at the same time, is

$$\frac{4}{3}\left(\text{FDT of single system}\right)z = \frac{4}{3}(0.0064)^2$$
$$= 5.5 \times 10^{-5} \text{ per year}$$

$$\text{The hazard rate} \qquad = \text{Demand rate} \times \text{FDT}$$
$$= 0.2 \times 5.5 \times 10^{-5}$$
$$= 1.1 \times 10^{-5} \text{ per year or once in } 90,000 \text{ years.}$$

Belts and Braces

The argument of the preceding section can be illustrated by considering belts and braces. The accident we wish to prevent is our trousers falling down and injuring our self-esteem.

Braces are liable to break and the protection they give is not considered adequate. Assume that breakage through wear and tear is prevented by regular inspection and replacement and that we are concerned only with failure due to faults in manufacture which cannot be detected beforehand and which are random events.

Experience shows that on average each pair of braces breaks after 10 years' service. Experience also shows that belts fail in the same way and as frequently as braces. Collapse of our trousers once in 10 years is not considered acceptable.

How often will a belt and braces fail together? If one fails then this will not be detected until it is removed at the end of the day. Assuming it is worn for 16 hours per day, then on average every man is wearing a broken belt for 8 hours every 10 years and broken braces for 8 hours every 10 years. The FDT of the braces is

$$\frac{8}{16} \times \frac{1}{10} \times \frac{1}{365} = 0.000137$$ and the FDT of the belt is the same.

The chance of the second protective device failing while the first one is 'dead' is:
$$\text{Hazard rate} = \text{Demand rate} \times \text{FDT}$$

$$= 2 \times \frac{1}{10} \times 0.000137 = 7.24 \times 10^{-5}$$ per year or once in 36,500 years.

Failure of belt and braces together, therefore, occurs once in 36,500 years. At the individual level this risk is acceptable. However, there are about 25,000,000 men in Great Britian so that even if every man wears belt and braces, 685 men will lose their trousers every year. At the national level it is considerable unacceptable that so many men should be embarrassed in this way.

To reduce the risk further, every man could wear a third protective device, a second pair of braces. This would reduce the failure rate for the individual man to once in 133,000,000 years† and for the country as a whole to once in five years. A third protective device, however, involves considerable extra capital expenditure and makes the system so complicated that people will fail to use it. An alternative is to get every man to inspect his belt and braces every 2 hours to see if either has broken. This will reduce the failure rate for the individual to once in $36,500 \times 8 = 292,000$ years and for the country as a whole to $685/8 = 85$ men per year. This may be considered acceptable but is it possible to persuade men to inspect their "protective systems" with the necessary regularity and what would it cost in education to persuade them to do so?

This example illustrates the following general points:

1   The risk can be reduced to any desired level by duplication of protective equipment but it cannot be completely eliminated. Some slight risk always remains. Even with three protective devices it could happen that coincident failure occurs not after 133,000,000 years, but next year.

2   The *method* used above is sound but the result is only as good as the *input data*. If the failure rate for belt or braces is not once in 10 years but once in 5 or 20 years, then the conclusion will be in error, not by a factor of two, but by a factor of four for two protective devices and by a factor of eight for three protective devices.

3 The event which we wish to prevent is not collapse of our trousers but injury to our self-esteem. Half (say) of the collapses will occur when we are alone or at home and will not matter, thus introducing an extra factor of two. (It is not explosions we wish to prevent but the damage and injury they cause; explosions which produce neither are acceptable.)

4 A risk which is acceptable to an individual may not be acceptable to the community as a whole.

5 It is easier to devise protective equipment or systems than to persuade people to use them. More accidents result from a failure to use equipment properly than from faults in the equipment. The high illegitimate birth rate, for example, is not due to failure of the 'protective equipment' but to the failure of the 'operators', through ignorance, unpreparedness or deliberate choice to use the equipment and methods available.

---

† Coincident failure or belt and two pairs of braces can occur in three ways, namely (a) belt fails when both pairs of braces have already failed, (b) braces 1 fail when belt and braces 2 have already failed, (c) braces 2 fail when belt and braces 1 have already failed. The FDT for a 1-out-of-2 system is $1/3\,f^2 T^2$ where f = failure rate (0.1 per year) and T = test interval (1/365 year).

For each failure mode the Hazard rate = Demand rate $\times$ FDT = $0.1 \times 1/3\,f^2 T^2$.

Hence: Total hazard rate $= 3 \times 0.1 \times 1/3\;f^2 T^2$

$$= 0.1\left(\frac{0.1}{365}\right)$$

$$= 7.5 \times 10^9 \text{ per year once in every } 133,000,000 \text{ years}$$

Literature cited

1   Lawley, H.G., *Chem.Eng.Prog.*, 70 (4), 45 (April 1974).
2   Lawley, H.G., *Hydrocarbon Process.*, 247 (April 1974).
3   ICI Ltd., *Assessing Projects*: Book 5, Risk Analysis, Methuen, London, 1968.
4   Kletz, T.A., *Major Loss Prevention in the Process Industries*, Inst. Chem. Eng. Symposium Series No.34, 1971, p.75.
5   Kletz, T.A., *Loss Prev.*, 6,15 (1972)
6   Bullock, B.C., *Chemical Process – Hazards with Special Reference to Plant Design – V*, Inst. Chem. Eng. Symposium Series No.39a, 1975, p.209.
7   Kletz, T.A., *Proceedings of the World Congress of Chemical Engineering, Chemical Engineering in a Changing World*, Elsevier, p.397, 1976.
8   Gibson, S.B., *Chem. Eng. Prog.*, 72 (2), 59 (1976)
9   Dicken, A.N.A., *The Quantitative Assessment of Chlorine Emission Hazards. Proceedings of the Chlorine Bicentennial Symposium*, p.244.
10 Sowby, F.D., *Health Phys.*, 11, 879 (1965).
11 Pochin, E.E., *Br. Med. Bull.*, 184, (1975)
12 Kletz, T.A., *Chem. Process.*, 77, (September 1974).
13 Kletz, T.A., and Lawley, H.G., *Chem Eng.*, 81,(May 12 1975).
14 Starr, C., *Science*, 165,1232 (1969)
15 Starr, C., in *Perspectives of Benefit – Risk Decision Making*, National Academy of Engineering, Washington, 1972, p.17.
16 Wall, I.B., Private communication
17 Turkenberg, W.C., *De Ingenieur*, 86 (10), 189 (1974).
18 Reed. J.D., *Loss Prevention and Safety Promotion in the Process Industries*, Elsevier, Amsterdam, 1974, p.191.

9 When the turning gets tough ...

Kerry Spackman and Sze Tan

Mathematics shows there is an ideal way to corner in motor racing. Only the drivers who apply this technique instinctively will become champions.

This fascinating article compares the route and speed with which a professional racing driver takes a bend, with the theoretically best route as determined by a mathematical model. You can see that a top class racing driver intuitively takes a route close to the best route, while an average driver follows a quite different line.

Getting to the top in motor racing requires the ability to drive a car consistently close to the performance limits of its tyres, engine and chassis: the sport ruthlessly weeds out those who cannot raise their car-control skills to the highest levels. But a driver who reaches the top international level (such as Formula One or Indycar racing) finds that most competitors possess very similar abilities. So what marks out the champions?

Virtually anyone can take a car to its limits on a straight track. Most races are won and lost where the cars are moving slowest – at the corners. The skill comes in choosing a speed and path that loses the least time negotiating them. This is where champions show their mettle.

Figure 9.1 shows a simplified "performance boundary curve" for one possible car and corner combination at a particular speed. Representing all possible speeds would require a family of similar curves or a three-dimensional picture. A skilled driver taking a corner slows down (applies longtitudinal deceleration), corners

For one car taking a corner, the limits of braking, cornering and acceleration are principally determined by the tyres. For a car driven at its limits around a corner, the car's acceleration follows a path from the bottom left of the diagram (braking), along to the right (turning the corner) and up to the top left (accelerating away)

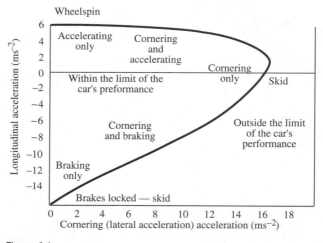

Performance boundary

Figure 9.1

(applies lateral acceleration) and then accelerates out and away (applies longitudinal acceleration). To control the car, the driver must ensure that the forces acting on the tyres are within the limits of their traction, and this involves a trade-off between the forces of braking and turning. As the driver corners, increasing lateral acceleration, the available longitudinal acceleration falls: high acceleration while turning too sharply causes a skid.

Top international drivers keep their cars just inside their performance boundaries almost all the time, though the boundary curve varies depending on the track conditions and the state of the car. Aerodynamic design complicates matters further. Air rushing over aerofoils on the car produces downforce that helps anchor it to the track. This allows greater cornering acceleration, but also increases wind resistance. The ratio of longitudinal to lateral acceleration alters as speed changes: at higher speeds the boundary curve becomes more flattened at the top and extends further to the right. Much of the skill that drivers acquire up to the international level lies in being able to read these changing variables, determine from moment to moment where the performance boundary lies and drive the car as close to it as possible. Acquiring this skill is a life's work, and very few drivers master it completely.

But for the elite who do, what is left to learn? Can anything help them further? Our studies with world champions indicate that one of the most important extra skills to learn is the optimum time to spend at the different parts of the performance boundary.

Figure 9.2 shows a similar curve to Figure 9.1, with bars added to indicate the amount of time spent at each part of the curve. Taking the corner, the [black] driver spends almost all of the time in the pure cornering region, at almost constant speed, or zero longitudinal acceleration. The [shaded] driver spends more time braking and accelerating. Both operate the car at its limit all the time.

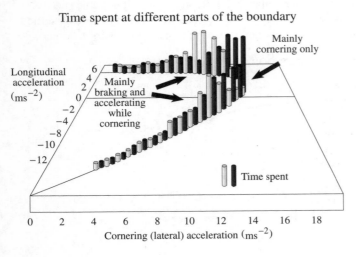

Figure 9.2

The bars in black indicate a driver taking the corner at an even speed; those in grey, one who brakes harder and accelerates away

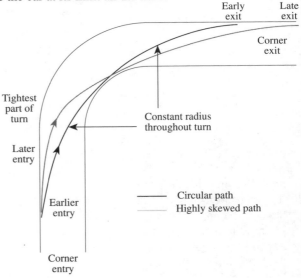

Figure 9.3

When translated onto the road, the black and grey bars would produce cornering paths like these. The grey path is more effective

We have developed computer simulations which show that for each boundary shape there is an optimum amount of time that should be spent on each part of the curve. The simulations take account of track variables, including the lengths of the straights before and after the corner, the angle the corner turns through, its inside and outside radii, and the surface's frictional coefficient. They also have to deal with vehicle variables, which are much more complex because they are interrelated: the cornering acceleration available at each speed depends on how hard the driver is braking and the line the car is taking. Cornering will cause the car to roll, and braking while cornering may cause it to yaw – so that it no longer points in the direction it is travelling. These factors alter the downforce, and thus the degree of cornering acceleration that can be applied before a skid occurs. From data acquired in repeated runs, we have built up matrices to describe the boundaries for cars under various conditions, and have fed these into the simulation.

The optimal solution differs from corner to corner, and from car to car on the same corner. What is clear is that a world champion will usually adopt a driving pattern that matches our computer simulations, whereas less skilled drivers consistently do not. This appears to distinguish true champions from other high-ranking drivers. More surprising is that, regardless of experience, most drivers never master this skill – and remain unaware of its importance.

Tighter routes to results

A driver can adjust the time a car spends on each part of the boundary curve by taking different paths through a corner. Figure 9.3 shows two possibilities. The [solid] line is a completely even, circular path. The [dashed] line is a very skewed path, in which the driver has started to turn later. The turn quickly becomes tighter than the smooth curve, but the driver exists in a wider curve further down the following straight. (There are of course infinitely more possible paths.)

These two cornering paths can be defined by their radii of curvature at each point, and this can be plotted to highlight the differences between the two paths. Tiny differences in paths are better highlighted by plotting the inverse of the radius. In Figure 9.4, the [solid] line shows this profile for the circular path, and the [dashed]

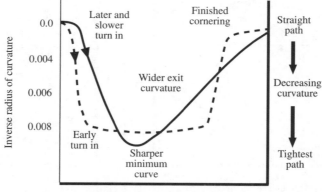

Figure 9.4

The radius of curvature indicates how sharply the car is turning; plotting the inverse high-lights the significant but small differences between the [solid] and [dashed] paths of Figure 9.3

line that for the skewed path. A driver following the circular path spends almost all the time at the "cornering only" region of the performance boundary, keeping the speed almost consistant throughout the corner. The tighter minimum radius of curvature of the skewed path forces the driver to enter the corner more slowly, but the gentler exit path allows a higher exit speed. Because the[dashed] driver changes speed more while passing through the corner, these differences are reflected in the time bars being spread more evenly round the boundary curve; the [solid] driver's time bars are more clustered.

In reality, most drivers take very similar paths through a corner. The differences are so small – typically around 10 centimetres – that they usually go unnoticed by even the keenest observer. To measure the differences we have developed a radio tracking system which can determine the position of the vehicle to an accuracy of 1 centimetre every hundredth of a second around an entire race track. It works by measuring the phase of radio reflections at receivers located around the track. Unscrambling the multiple reflections including those from obstacles around the track is a difficult task that requires a sophisticated computer program and parallel-processing transputers. Ford is now using this system for car development.

Using our location system we compared the paths taken by two drivers round one corner of a Ford test track. One was Jackie Stewart, a three-time Formula One world champion now aged 52. "Driver B" was a highly competitive 24-year-old European driver now in Formula 3000 racing, the level below Formula One. Figure 9.5a shows the curvature profiles measured on three passes through the corner by Stewart. Figure 9.5b shows two passes of the same corner by driver B in the same car.

There are two obvious differences. First, Driver B's profiles show lots of wobbles and inconsistencies, while Stewart's are extremely smooth and consistent. More important is the overall shape of the plots.

The corner has an inner radius of 150 metres and an angle of turn of 85°. The inside edge is therefore at least 223 metres long, though the drivers follow a wider and therefore longer path. Driver B's curvature profile is much flatter than Stewart's:

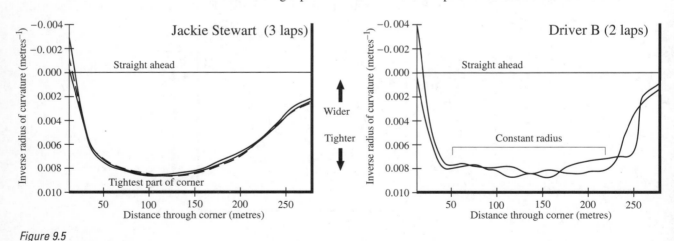

Figure 9.5

A comparison like Figure 9.4 of the paths taken in the same car of the same corner by Jackie Stewart, the ex-world champion, and a professional racer from Formula 3000. Stewar's resemble the [dashed] curves from Figure 9.3 and 9.4; Driver B's are more like the black curves

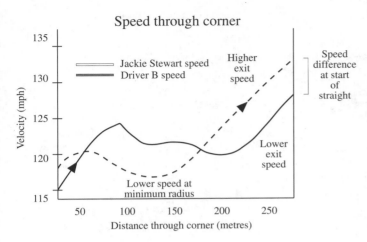

Speed through corner

Figure 9.6

Although Drive B has a higher speed throughout the corner, Jackie Stewart accelerates out of it more effectively and has a higher speed going into the following straight

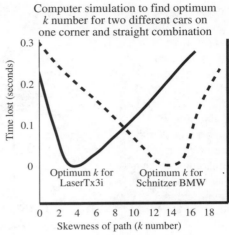

Computer simulation to find optimum *k* number for two different cars on one corner and straight combination

Figure 9.7

Even on the same corner, different cars have different "best" lines. The Ford Laser Tx3i, based on the Ford Escort, must be driven on a gentler curve than the racing BMW

the radius of his path 50 metres into the corner remains almost unchanged until after 230 metres. The bottom of his profile is almost square. Stewart's profile is much more rounded, closing up to a tighter entry curve about 100 metres into the corner and then straightening out to almost twice the minimum radius after 230 metres. Driver B's path is like the [solid] line in Figure 9.4, while Stewart's resembles the [dashed] line.

Stewart actually spends more time in the corner than Driver B because he has to take the sharper curve more slowly. However, he can exit faster because of his wider finishing radius, and this higher speed advantage stays with him all the way down the following straight, more than making up for time lost in the corner. Taking the corner and the straights on either side as a single problem, Stewart found the fastest solution (see Figure 9.6).

Choosing how much to skew the corner is a very complex problem. Skewing too much (that is, turning too late) means that the corner must be taken so slowly that the time lost there cannot be recouped fully in the following straight; it may even result in a lower exit speed. Skewing too little (turning early), like Driver B, can also lose speed in the straight.

We have developed a measurement that we call the "*k*" number (see the section Nigel Mansell does this in his head – can you?), which describes how skewed a curve the driver has taken through the corner. A path which is completely circular throughout has a *k* of 0. Higher values of *k* describe increasingly skewed paths. Specifically, *k* refers to how quickly the path flattens out. Our computer simulations try different values of *k*, then look at the paths that result for various car boundary performances and work out the time taken for the corner. The simulations show that there is no single optimum value of *k* for all cars or corners: a car taking different corners will require different solutions. Even for the same corner, different cars will require different values of *k*. This is shown in Figure 9.7, which plots *k* against the time taken by two cars, a BMW M series set up by Schnitzer (The

Drivers who develop their skills on a level track (left diagram) gauge when to accelerate into a corner from the ratio of the inertial (centrifugal) force F to their weight, W. But on a banked track (right diagram), the inertial acceleration F' that the driver perceives is reduced, while the perceived weight W' increases. The unaware driver under-estimates the lateral acceleration and turns into the corner too late, missing the best line

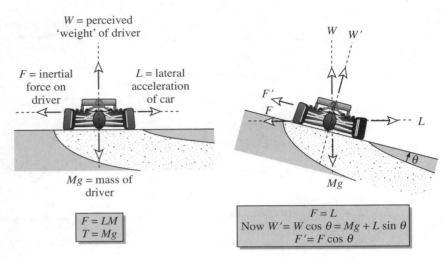

W = perceived 'weight' of driver

F = inertial force on driver

L = lateral acceleration of car

Mg = mass of driver

$$F = LM$$
$$T = Mg$$

W W'

F'
F

L

Mg

θ

$$F = L$$
$$\text{Now } W' = W \cos\theta = Mg + L\sin\theta$$
$$F' = F\cos\theta$$

Figure 9.8

World Group A Saloon car champion works team) and a Ford Laser Tx3i (a 1.8 litre fuel-injected model derived from the Ford Escort) to complete a single corner and the two straights on either side. The minimum time for the BMW occurred at k = 13.8. For the Laser Tx3i it was closer to $k = 4$, which means for best results in this corner the Laser should be driven in a much rounder curve than the Schnitzer.

In general, the shorter the corner and the longer the straights either side, the more the path should be skewed. Similarly, the greater the ratio of the car's potential forward acceleration to cornering acceleration – a measure defined by the car's speed and performance characteristics, not the corner – the more the path should be skewed.

We have measured numerous drivers in a range of cars over four years in Britain, the US, Germany, Japan and New Zealand, and found that only champions uniquely seem to possess the ability to approach the correct k number, although even they are not perfect. Lesser drivers seem to fall into certain stereotyped patterns, and they fail to adjust their driving appropriately for different conditions – even though they still drive the car at the boundary limit. This explains why some drivers can be expert in one type of car yet struggle in a different class. The true champion, on the other hand, can quickly approach the optimum k for any car. For example, on our test track Stewart consistently drove a slightly different line with a Ford Mustang than with a Ford Thunderbird; these differences are predicted in our computer model.

Looking at data collected from accelerometers on the cars driven by Stewart and Driver B, it is Driver B who turns out to have had slightly higher cornering accelerations throughout the entire curve. As a younger driver, with slightly faster reflexes, he can drive the car slightly closer to the limit than the retired Stewart. However, Stewart is still quicker because he selects better k values. Driver B called on tremendous car-control skills to take the car closer to its performance limits. The trouble is, they were the wrong parts of the limits, and he pushed the car so hard that he had difficulty controlling it. This is what caused the wobbles in his curves (Figure 9.5b). Information derived from our computer simulation could make up for the sensitivity to car and track that all but the best drivers lack. By debriefing drivers after practice circuits, or giving them instructions by radio, it should be possible to train them to choose a better k under different conditions.

'Formula One drivers who switch to Indy racing find the banked tracks make them turn into the corners too late. One needed a year to adjust'

So what do our findings imply for Nigel Mansell, transferring from Formula One to Indycar racing? Among other differences, most Indy tracks have steeper banking – up to 9°12' – than is usual in Formula One. Cornering on a slope means that part of the lateral acceleration acts in the direction that the driver perceives as downwards (Figure 9.8), and this can fool people used to level tracks into underestimating the ratio of lateral to longitudinal acceleration acting on the tyres. As a result, they may try to corner too late and too sharply. Mansell will have to learn how to interpret these differently perceived forces to find the optimum k: at 200 mph on a quarter-mile corner (as at the Indianapolis 500 racetrack), the banking creates a 2 per cent different in the perceived ratio of lateral to longitudinal acceleration – small, but significant at those speeds.

Other drivers who have switched from Formula One to Indycar racing confirm that this is a real problem. Initially they consistently overestimate the required k – that is, they turn too late. It takes time to learn to turn in earlier, to take the more even line that a banked track requires. The problem is further compounded by the different aerodynamics of Indycar and Formula One designs. Even drivers supposed to be at the peak of their abilities find the change confusing. One who made the switch needed a year to adjust.

Unfortunately, finding the optimum solution is nowhere near as simple as just looking at the acceleration ratios as we have done here. There are many other variables, such a tyre slip (in which the tyres slide, minutely, during cornering without the car skidding out of control). Also, complex interactions between the throttle and a car's steering have an important effect which differs from car to car. However, we believe that the next revolution in motor racing could well come from using sophisticated mathematical analysis and real-time feedback techniques to give drivers a competitive advantage. Military pilots already rely on real-time computer analysis and feedback from head-up displays to assist them in combat. Applying many of those techniques, together with the type of mathematical treatment outlined above, will certainly make as big a contribution as standard telemetry analysis of the car itself has already made.

Nigel Mansell does this in his head – can you?

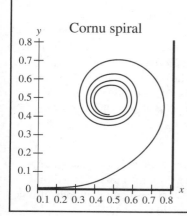

Cornu spiral

Deciding how late to turn into a corner is a matter of getting the right value for k. We start with the cornu spiral (a curve whose curva-ture K at any point is proportional to the distance from the zero point). This spiral can be generated from the Fresnel integral, usually used for analysing intensities of diffraction patterns, which gener-ates two variables x and y from the value of a third u. Each value of u gives a different value of x and y when the pair of integrals generat-ing them are solved (seeequations).

$$y = \int_0^u \cos\left(\frac{\pi v^2}{2}\right) dv \qquad x = \int_0^u \sin\left(\frac{\pi v^2}{2}\right) dv$$

We then look at the path taken by the driver, analyse the small section around its tightest radius of curvature, and scale it to the cornu spiral. Then we slide the driver's curve along the spiral, looking for the point where the two curves match most closely. This happens at a particular value of the Fresnel integral's top value u. The k value is then the inverse of u.

Of course, we do this on a computer after exhaustive measurements on the track. The drivers have to find the best k using just their senses and instinct. Perhaps it's not surprising that so few of them truly master the ability to find the right line.

10 Arithmetic in the cradle

P E Bryant

Read this short article while you are working on the *Distributions* unit in Book 5. While you read it, think about how would you design an experiment to verify the results which the author obtained. What would your null hypothesis be? And your alternative hypothesis? And what data would you collect?

Babies can apparently make simple arithmetical calculations, according to a series of experiments described by Karen Wynn[1]. The appearance of this paper is a notable event in the history of development psychology – Wynn claims that babies as young as five months in age are able to add and subtract, and she concludes that the basis of arithmetical understanding may be innate. Her results and her conclusion are surprising, even in an area of research which has already produced many surprises.

Over the past 20 years we have learned a great deal about the capacities of very young babies. Well within their first six months, we now know, babies can tell objects apart by their shape[2], size[3] and colour[4]: they know that these objects go on existing when they are hidden[5] and they also can take in whether they are solid or not[6]. Babies can even tell, when they listen to someone talking, whether that person's lip movements are the appropriate ones for the speech that they are hearing[7,8]. These are striking skills to find in a creature which used to be thought of as utterly imcompetent and ineffective, but they could all in one way or another be described – even dismissed – as perceptual. One could argue in the face of this evidence that the perceptual system of very young babies is intact, so that they can take in information about their physical surrounds, but that they may not be able to think or reason in any way.

The distinction has begun to look implausible, however, for there is now considerable evidence that children can recognise the number of a set of objects and can tell when there has been a change in that number – at any rate when small numbers are involved – and that they can do so irrespective of changes in the spatial arrangement of the set. They can even tell whether the number of drumbeats that is played to them coincides with the number of objects in a visual display that is shown to them at the same time[9-11]. To recognise that two sets have the same number despite radical differences in their perceptual appearance is to go well beyond simple perception.

The experiments described in this issue provide remarkable evidence that young babies' intellectual skills may go a good deal further even than this. Not only do they take in information about the number of objects before them: they may also be able to work out the results of an addition or a subtraction.

Wynn used a technique which has proved extremely useful in the past. Babies tend to look for a relatively long time at displays that are new or unexpected to them. In her first two experiments the babies saw a hand putting one or two objects on a table and then a screen being raised in front of the object(s). Then they saw the same hand either adding an object to the one behind the screen (1+1) or taking one away (2 − 1) from behind it. Finally the screen was lowered, and the baby could see how many objects there were now on the table. Half the time that number of objects was as it should have been, but the other half it was not. In the latter case, the babies saw two objects when there should have been one, or one object when there should have been two.

Wynn reports that the babies looked significantly longer at these incorrect displays than at the correct ones and argues from this that they had anticipated the correct number of objects and were surprised when they saw a different number. To do so, she claims, they must have carried out some form of mental arithmetic.

Her final experiment tackles a possible objection to this conclusion. In the first two experiments the number of objects in the incorrect trials was the same as their number before the addition or subtraction: so the babies might simply have realized that the final number of objects would have to be different from the original number, and therefore were surprised after the screen was lifted to see no change. In the third experiment Wynn took care of this objection by giving them an addition (1+1) in the same way as before, but arranging things so that after the screen was lifted they either saw two objects there (the correct display) or three. Thus the number of objects in the incorrect display was not the same as the number before the addition. Yet the babies still looked at the incorrect displays for a longer time than they looked at the correct ones.

The control is a convincing one, though it should be added that Wynn's argument would have been stronger if she had put in the same control for subtraction as well. As it is, she has made a convincing case only for babies of five months being able to perform addition. It is quite easy to think of an experiment to establish as convincing a case about subtraction.

The apparently cast-iron evidence that Wynn offers us to show that very young babies can carry out additions in their head establishes that they are capable of arithmetiçal reasoning very early on in life. But this prompts some further questions. One is whether young babies have a full understanding of addition and subtraction. In my view Jean Piaget was correct when he argued that someone who can add or subtract correctly does not necessarily understand addition and subtraction[12]. One has to grasp the relationship between these two mathematical operations in order to understand either of them properly. Piaget claimed that children do not realize the inverse relationship between addition and subtraction until they are roughly eight years old, and only then as a result are able to learn about the additive composition of numbers (for example, that if 5 and 3 add up to 8, then 8 − 5 must equal 3). This is an idea that has never been challenged empirically, but Wynn's work suggests that it would be quite easy, and well worth the trouble, to devise an experiment to see whether babies are able to understand the inverse relationship between addition and subtraction.

Another very different question concerns the relationship between the early skills that Wynn has discovered in babies and the progress that they make in mathematics later on. Wynn speculates that babies' evident arithmetical capacity "may provide the foundations for the development of future arithmetical knowledge". The suggestion may seem extravagant, but it is really quite a plausible one, given the results of some longitudinal work which shows that individual babies' performance in other experiments involving measures of looking predicts their abilities later on reasonably well. Several studies[13,14] have established that looking time in experiments on babies' perceptual skills is strongly related to their later IQ. It is a curious relationship, given the simplicity of the earlier measures and the very general nature of the IQ test.

Wynn's hypothesis suggests that one should try to find a more specific and therefore more satisfying connection. It seems quite possible to me that her early measure of babies' addition might be strongly related to their later success in mathematics but not, say, in learning to read. Other measures, possibly to do with sound discrimination, might predict their later success in reading but not in mathematics. Wynn's interesting speculation that she may have uncovered the foundations of arithmetical knowledge should remind us that we can no longer be content with discovering surprising skills in very young babies. We must explore the significance of these skills in children's lives later on.

References

1   Wynn, K. *Nature* **358**, 749–750 (1992)

2   Slater, A. Morison,V & Rose, D. *Br J. dev. Psychol.***1.** 135–142 (1983)

3   Slater, A.& Morison ,V. *Perception* **14**, 337-344 (1985)

4   Adams, R.J., Maurer, D. & Davis, M. J. *Exp.Psychol* **41**, 267–281 (1986)

5   Baillargeon, R., Spelke, S & Wasserman, S. *Cognition* **20**, 191-208 (1985)

6   Baillargeon, R. *Cog.Dev* **2** 179–200 (1987)

7   Dodd, B. *Cog. Psychol.***11**, 478-584 (1979)

8   Kuhl, P.K. & Meltzoff, A.N. *Science* **218**, 1138–1141 (1982)

9   Antell, S.E. & Keating, D.P. *Child Dev.* **54,** 695–701 (1983)

10   Starkey, P. & Cooper, R. *Science* **210**, 1033–1034 (1980)

11   Starkey, P., Spelke, E.S. & Gelman, R. *Cognition* **36**, 97–128 (1990)

12   Piaget, J. *The Child's Conception of Number* (Routledge & Kegan Paul, London, 1952)

13   Bornstein, M.H. & Sigman, M.D. *Child Dev.* **57** 251-274 (1986)

14   Rose, D. & Slater, A. *Intelligence* **10**, 251-263 (1986)